Chinese "Cancer Villages"

AF736505

Chinese "Cancer Villages"

Rural Development, Environmental Change and Public Health

Chen Ajiang,
Cheng Pengli,
and Luo Yajuan

Translated by
Jennifer Holdaway, Qi Di and Wei Han

Routledge
Taylor & Francis Group
LONDON AND NEW YORK

Chinese "Cancer Villages": Rural Development, Environmental Change and Public Health was originally published in Chinese by Chen Ajiang, Cheng Pengli and Luo Yajuan, China Social Sciences Press, Beijing, 2013.

First published in 2020 by Amsterdam University Press Ltd.

Published 2025 by Routledge
4 Park Square, Milton Park, Abingdon, Oxon OX14 4RN
605 Third Avenue, New York, NY 10158

Routledge is an imprint of the Taylor & Francis Group, an informa business

© Chen Ajiang, Cheng Pengli & Luo Yajuan / Taylor & Francis Group 2020

All rights reserved. No part of this book may be reprinted or reproduced or utilised in any form or by any electronic, mechanical, or other means, now known or hereafter invented, including photocopying and recording, or in any information storage or retrieval system, without permission in writing from the publishers.

Trademark notice: Product or corporate names may be trademarks or registered trademarks, and are used only for identification and explanation without intent to infringe.

ISBN: 9789089647221 (hbk)
ISBN: 9781041176794 (pbk)
ISBN: 9781003692331 (ebk)
NUR 740

Cover photo: Pollution from metal mining flows through the countryside
Source: Cheng Pengli, 2010

Cover design: Coördesign, Leiden

DOI: 10.5117/9789089647221

Every effort has been made to obtain permission to use all copyrighted illustrations reproduced in this book. Nonetheless, whosoever believes to have rights to this material is advised to contact the publisher.

For Product Safety Concerns and Information please contact our EU representative:
GPSR@taylorandfrancis.com
Taylor & Francis Verlag GmbH, Kaufingerstraße 24, 80331 München, Germany

Table of Contents

List of Figures, Graphs and Tables

Acknowledgements

This book is the result of our many years of research on 'cancer villages'. This research was supported by two grants from the Social Science Research Council's China Environment and Health Initiative for the project 'Perceptions and Response Strategies to Environmental Health Risks' (RBF/SSRC-CEHI/2010-03-01; RBF/SSRC-CEHI/2011-04-02), and a China National Social Science Foundation Grant for the project, 'Harmony between People and Water: based on a field study of Lake Tai and the Huai River Basin' (07BSH036).

The completion of this book was assisted by the interdisciplinary Forum on Health, Environment and Development. Co-Directors of FORHEAD Dr Jennifer Holdaway (initially at the Social Science Research Council, later at Oxford University and currently with the International Institute for Asian Studies, University of Leiden) and Professor Wang Wuyi (Institute of Geographical Sciences and Natural Resources Research) organized several meetings at which researchers from different disciplines were able to interact and discuss this fascinating but extremely challenging research topic. Dr Holdaway and Professor Wang wrote a preface for the book, which encouraged us and spurred us on. We learned a great deal from our exchanges with many members of the FORHEAD network including Zhang Shiqiu (Peking University), Yang Linsheng of the (Institute of Geography of the Chinese Academy of Sciences) and Su Yang (Development Research Centre of the State Council), and we would like to express our sincere thanks.

Dr Jennifer Holdaway reviewed and edited the translation of every chapter and translated the introduction. In this process, she came across some gaps in the original Chinese and suggested adjustments to make the narrative more accessible to English language readers. We do not have the words to thank her for taking time out of her busy schedule to do this time-consuming work.

We are very grateful to Qi Di and Wei Han for their hard work in completing the first round of translation.

Special thanks to Professor Zhou Yejin, Dr Chen Tao, Dr Wang Sitong, Wang Yanqing, Zhou Dun, Song Liangguang and Zhou Yan, who took part in some of the fieldwork.

Many people helped us in many ways in the process of fieldwork, including but not limited to village cadres and doctors, villagers themselves, enterprise workers, township cadres and officials working in county and municipal Environmental Protection and Water Resources agencies and Centres for Disease Control and Protection. We regret that these people are too numerous for us list them all individually by name, but we would like to express our heartfelt thanks.

Preface to the English Language Edition

Jennifer Holdaway and Wang Wuyi[1]

We are delighted that this book is finally seeing the light of day in the English language. As some readers are aware, this has been a long and difficult process, and we are grateful to Amsterdam University Press (AUP) for its support.

Several of the chapters in this book first appeared as journal articles, and in 2013 they were collected in a Chinese language volume, titled 'Cancer Village Research'. In 2014, Professor Chen and his colleagues signed a contract with AUP for the English version of the book containing additional material, and in 2016 we expanded our original preface to include a discussion of the new material. Although several years have passed, we have chosen to retain that 2016 version here and add just a short coda for the English edition, so that the evolution of the content of the book and our thinking about the topic is clear.[2]

Preface to the 2016 Edition

'Cancer villages' are a highly emotive issue. Cancer is a 'dread disease,' more frightening to most people than other equally life-threatening illnesses, such

1 The authors are Co-Directors of the Forum on Health, Environment and Development (www.forhead.org). Jennifer Holdaway is currently an Affiliated Fellow with the International Institute for Asian Studies (IIAS) of the University of Leiden and a Foreign Expert at the Institute of Geographic Sciences and Natural Resources Research of the Chinese Academy of Sciences. She was formerly a Senior Research Fellow with the School of. Global and Area Studies at Oxford University and prior to that, Program Director and China Representative at the Social Science Research Council. Wang Wuyi is Professor and Former Deputy Director at the Institute of Geographic Sciences and Natural Resources Research of the Chinese Academy of Sciences.

2 Where available, we have replaced Chinese language references with English language versions of the same publication, and conference papers with published versions of the same text. Aside from the discussion of new material and new relevant publications, this text is only minimally different from the preface to the 2013 Chinese language book.

Chen Ajiang, Cheng Pengli, and Luo Yajuan, *Chinese "Cancer Villages": Rural Development, Environmental Change and Public Health*. Taylor & Francis Group 2020
DOI: 10.5117/9789089647221_PREF

as heart or respiratory failure. Since villages generally conjure up images of a natural and healthy rural environment, the idea of their being plagued by cancer is therefore especially unexpected and disturbing. Although poverty has decreased dramatically and the social safety net has expanded in recent years, overall, it is widely acknowledged that rural Chinese have also received fewer benefits from rapid economic development than urban dwellers. Reports of cancer villages suggest that they may also be bearing the brunt of its negative environmental impacts.

Media reports of cancer villages are shocking, citing rapid rises in cancer deaths and showing photographs or film of skeletal victims and grieving families. But how do media accounts and public perceptions of 'cancer villages' relate to what we know about the phenomenon from research; and how can research help to address the problem? This book explores these questions.

Although the government has acknowledged that 'cancer villages' exist, little is known about them.[3] In the aggregate, cancer rates in China have been rising for some time and data from the Ministry of Health shows that the disease is now the leading cause of death in rural as well as urban areas.[4] Many cancers have a long latency period and the increase in its frequency can be seen to broadly follow the curve of China's economic growth. To a certain extent, this is just part of a larger 'epidemiological transition' resulting from longer life spans, and lower mortality from communicable diseases associated with poverty as well as improved diagnosis and access to health care. It also reflects changes in diet and lifestyle that contribute to cancer, cardiovascular illnesses, and other 'diseases of affluence'.[5] In fact, when the ageing of the population is taken into account, mortality rates for some cancers (including some types of liver cancer, stomach and oesophagus cancer) have fallen between 1990 and 2013, while others (liver cancer due to hepatitis C, prostate cancer and pancreatic cancer) have shown an increase.[6]

There is little doubt that part of the increase in certain types of cancer is due to environmental pollution. But how much, and the contribution of pollution to disease in any particular location, is hard to determine. The carcinogenic effect of certain chemicals is well understood on a biological

3 Ministry of Environmental Protection, 'Huaxuepin fangkong shierwu guihua'.

4 Ministry of Health, '2012 nian woguo weisheng tongji tiyao'.

5 Gong et al., 'Urbanization and Health in China.'; van de Poel et al., 'Is There a Health Penalty of China's Rapid Urbanization?'; World Bank. *Toward a Healthy and Harmonious Life in China.*

6 Zhou et al., 'Cause-Specific Mortality for 240 Cases in China During 1990-2013'.

level, and even without detailed information about specific pollutants, robust statistical correlations can be established between environmental quality and health for large populations. For example, the Atlas of the Water Environment and Cancer in the Huai River Valley has demonstrated a clear association between water pollution and high rates of digestive tract cancer at the county level using regular environmental monitoring data. The analysis shows that some counties which had higher than average rates of cancer in the 1970s have seen significant improvement, while others that were on the low end have seen a rise in cancer that is clearly associated with changes in water quality. The association remains even when lifestyle factors are controlled. As water quality has improved, cancer rates have also fallen, although the reasons for this are complex and include not only environmental protection but also improved health services and possibly nutrition.[7] Recent provincial level analysis of the causes of death also shows regional clustering in types of cancer and interesting patterns of change that provide the basis for understanding how different environmental conditions as well as dietary patterns and unequal access to healthcare may be shaping health outcomes.[8]

But although associations between environmental quality and cancer can be demonstrated on the level of large populations (in China usually the county), proving the existence of cancer clusters and attributing causality at the village level is extremely hard. Even when there are good records of the number of cancer deaths, too many other factors can be involved in the onset of disease, including genetics, personal behaviours such as diet, exercise, smoking and alcohol use, and general health status and nutrition. Often this information is lacking and even if it is not, samples are too small to distinguish the role of different factors. Although the lack of good and/or publicly available data on many of these variables in China exacerbates the problem, this difficulty is not unique to China. Attempts to prove the health effects of pollution on small populations have been the subject of bitterly contested and protracted law suits in the United States, and a substantial literature has documented efforts by communities to gather evidence and gain acknowledgment of these contested illnesses.[9] It is quite rare to find situations, like the Minamata case in Japan, where disease can be clearly linked to a particular pollution source; and even in that case, responsibility and compensation were contested for many years.

7 Yang and Zhuang, *Atlas of the Huai River Basin Water Environment: Digestive Cancer Mortality*.

8 Zhou et al., 'Cause-Specific Mortality for 240 Cases in China During 1990-2013'.

9 For example, Brown, *Toxic Exposures*; Tesh, *Uncertain Hazards*.

In China, the compressed nature of the industrial transition, and the multiple social changes that have accompanied it, often make it particularly challenging to prove environmental health effects at the village level. China's industrialization has been much more rapid than that of Europe, the United States, or even Japan. Furthermore, unlike many parts of the world, a lot of China's industry has historically been located in the countryside, and in many areas there has been rapid turnover in the type and scale of industrial activity. Even during the Mao era, Third Front and other policies promoted the establishment of industries in rural areas in China's interior, and from the early 1980s, in response to the policy incentives and market forces associated with reform and opening up, Town and Village Enterprises (TVEs) sprang up all over the country.[10] These rural industries and the household workshops that accompanied them were generally small, used basic technologies and released their waste directly into the surrounding atmosphere, water and soil.[11] Many later closed or changed their product lines or form of ownership, with the result that many of China's rural areas have been affected by waves of pollution from different sources.

So while there are cases in which mines or individual industries are the clear source of health problems, there are also many situations in which complex sources and forms of pollution make it very hard to establish causality. Further complications arise from the fact that the populations affected by pollution have not been stable. Rural-urban migration has profoundly changed patterns of residence and occupation, with many people who came of age in the reform era working and living in a number of places over the course of their lives, often in jobs that have exposed them to environmental health risks.[12] Meanwhile, improved communications and media, along with financial and social remittances associated with migration, have in many places blurred the boundaries between formerly distinctive urban and rural lifestyles and the diseases associated with them.[13] These changes in place of residence, occupation and lifestyle make it difficult to isolate the impact of exposure to pollution on health, and even harder to attribute responsibility to a particular source. Migration means that many villages also have a very uneven age distribution, with healthy adults often away working in cities, leaving older people and those in poor

10 Bramall, *The Industrialisation of Rural China*.

11 Tilt, 'The Political Ecology of Pollution Enforcement in China'; Wang et al., 'Rural Industries and Water Pollution in China'; Han and Zhang, 'China's Environmental Governance of Rapid Industrialization'.

12 Hu et al., 'Internal Migration and Health in China'.

13 Holdaway, 'Environment, Health and Migration: towards a more integrated analysis'..

health behind. Migrants who are very sick are also most likely to return.[14] Although every situation is different, overall, patterns of migration skew many rural populations in the direction of poor health.

The Chinese village is therefore a very difficult unit of analysis when it comes to studying patterns and trends in the burden of disease and in epidemiological research the county is usually the unit of analysis. At the same time, despite all these complexities, it is still clear that in a substantial number of villages mortality from cancer is much higher than the rural average and many of these villages also have a history of industrial pollution. Even if rigorous epidemiological analysis is impossible, the likelihood of an association seems high.

Furthermore, as Chen Ajiang and his colleagues argue in this book, the village remains a key social and administrative entity. Urbanization and demographic change may lead China's villages to lose their historical significance as the primary source of social identity and economic security within a few decades, but for now, they remain an important unit of social life. And so, regardless of whether the relationship between pollution and cancer can be established at this level, these rural communities will continue to be the focus of claims and contestations for some time to come. For all these reasons, Chen Ajiang is right to insist that 'cancer villages' are a phenomenon with which both policy and research must engage.

But it is a brave social scientist who enters this highly charged and complex territory, and there are hard choices to be made in how to deal with the scientific uncertainty surrounding the relationship between pollution and cancer. One option is to take media reports or citizens' claims about the causes of disease at face value and focus on their struggles to close factories or win compensation for health damages.[15] Another is to set the question of causality aside and study the landscape of risk perception and the ways in which different understandings shape behaviour.[16] Both of these are valid approaches from a social science perspective, and research along these lines has yielded valuable insights.

Chen Ajiang and his colleagues have taken the bolder step of tackling the problem of uncertain causality and its implications head on and engaging with natural and medical as well as social science data. This was a decision

14 Hu et al., 'Internal Migration and Health in China'.

15 For example, Deng and Yang 'Pollution and Protest: Environmental Mobilization in Context'; van Rooij 'The People vs. Pollution: Understanding Citizen Action against Pollution in China'.

16 Lora Wainwright et al., 'Learning to Live with Pollution'; Lora Wainwright, 'An Anthropology of "Cancer Villages"'; Jing, 'Environmental Protests in Rural China'.

for which Professor Chen received some criticism at an early stage of the project from epidemiologists who felt that assessing cause-effect relationships was not something that a social scientist should attempt. But he persisted, and his determination to understand and weigh the different sources of evidence regarding the relationship between pollution and disease, and consider the role of social scientists in working on these issues, has produced some of the most interesting insights in his research.

Over a period of many years, Chen Ajiang and his team have conducted extensive field research in villages in, Henan, Jiangsu, Zhejiang, Jiangxi, and Guangdong. The studies published in the first edition of this book in 2013 all focused on villages which had experienced significant levels of industrial pollution and had cancer rates at least twice the rural average, and sometimes much higher. In each case, though with different degrees of confidence and assertiveness, villagers attributed cancer to pollution from local industry or mining.

Instead of shying away from the difficult question of whether these claims were justified, the researchers assembled the available evidence to determine what can be known about the relationship between pollution and cancer. They examined all the information they could find on recent levels of pollution and health outcomes, including the results (where available) of tests conducted by government agencies, but also more informal data gathered by villagers. They did their best to understand the manufacturing processes used by local industries and the kinds of pollution they created. They also dug deep into the past, leading us through the history of these communities over the course of the last 30 years, tracing their different development trajectories and the implications of these for the environment, and for public health.

This multidimensional approach enabled them to distinguish between cases in which a clear relationship between pollution from a particular source (usually a single factory or mine) and cancer could be quite clearly established; those in which the evidence was quite strong but not conclusive; and other cases in which the evidence for blaming a particular pollution source proved either weak or non-existent. In the last case, although there were polluting industries in the area, the particular pollutants involved were not known carcinogens or the pollution was too recent to be the cause of cancers with long latency periods. Examination of the history of these places complicated the search for a culprit because a series of industries have often come and gone over the years. And in some cases, such as the Huai River Basin, waves of pollution from industrial accidents upstream from the village must be considered in addition to local pollution sources.

In these situations, although the pollution-disease relationship seems clear on a general level, no single specific industry can be implicated as the responsible party.

In addition to carefully unpacking the evidence with regard to pollution, another major contribution of these case studies is that they integrate analysis of the available environmental data with attention to individual and social factors that may be contributing to the onset of cancer and other diseases. In attempting to assess the role of pollution to cancer in Huangmengying village in Henan, Chen found that smoking and hepatitis partly explained the greater prevalence of lung and liver cancer among men. Local customs such as drinking 'raw' water (*shengshui*), which increased stomach infections, also emerged as possible contributing factors. Pursuing a theme present in his earlier research, he also argued that living with 'external' pollution from industry made villagers resigned to a dirty environment and less concerned about the impact of their own 'internal' pollution from household waste or agricultural activities.

This edition contains new material, including a chapter on Lianshui County in Jiangsu which focuses on interactions between water, agricultural practices and health. Epidemiological data show that rates of oesophagus cancer in the county are three times the national rural average, accounting for 34% of all deaths from cancer. Records show that the disease has always been regionally clustered, with northern Jiangsu, along with the Henan-Hebei border and parts of Sichuan being hotspots. There is no conclusive explanation, but some of the factors known to be associated with the disease, including nitrosamines and aflatoxins in food, protein deficiencies and certain dietary practices (such as consuming preserved food and very hot drinks) are also common in those places, where a number of studies have been conducted going back to the 1950s.

Drawing on historical records, Chen investigates the changes in the environment and rural livelihoods that might have contributed to greater risk of cancer through what epidemiologists have referred to as the 'nitrogen cycle hypothesis,' which suggests that high levels of nitrates in drinking water may combine with other substances in the body with to form carcinogenic *N*-nitroso compounds. He traces the serious impact on the county of changes in the course of the Yellow River, which caused flooding, water-logging and drought in the Huai River Basin. This left much of the land salinized, leading to low yields and a lot of land being left fallow. The county was desperately poor, with frequent famines and widespread malnourishment. These problems were successfully addressed by water engineering projects in the 1970s which irrigated the dry land and made it possible to replace low value coarse grains with paddy rice. Along with the introduction of high yield seeds and the use of chemical fertilizers and pesticides, this led to rapid increases in production and freed the region from hunger.

However, Chen hypothesizes that this reconfiguration of the water system may also have affected villagers' drinking water, with negative effects on health. He notes that villages in the region have the custom of digging long trenches (called 'river ponds' or 'big wells') near their homes which provided earth to raise their houses above the plain and prevent flooding, and were traditionally also used for drinking water and raising fish and shrimp. But the 1970s engineering works linked these pools up to irrigation systems and therefore exposed them to contamination by fertilizers and pesticides as well as manure from the burgeoning livestock raising industry, increasing the risk to health. This interpretation is further strengthened by the fact that improvement of drinking water sources in other regions has effectively lowered the oesophagus cancer rate; although similar records have not been kept in Lianshui County, lowered rates there may also reflect improvements in drinking water sources as the county has become wealthier.

From an epidemiological perspective, Chen's account is not conclusive. His data on village water sources and environmental change cannot prove the link to cancer; only that there is evidence of some of the factors that are associated with it in the scientific literature. However, it does raise avenues of enquiry that could usefully be followed up by medical science studies and which might contribute to a better understanding of the variation in patterns of cancer in the area and suggest ways to reduce risk.

In addition to examining the medical data and relevant environmental factors, the cases also document how villagers understand the relationship between pollution and illness. These include changes in the taste and smell of water and food, death or illness among animals, and observation of new types of illnesses among family and neighbours. As Anna Lora-Wainwright has done elsewhere, the authors note the way in which rural people deploy quite sophisticated 'lay epidemiologies' to understand the causes of disease,[17] in the context of the 'society of familiars' (*shuren shehui*) of Chinese village life. But they also note how low levels of education limit villagers' ability to evaluate information from scientific tests, and their tendency to suspect current and local pollution sources that are visible to the senses, even in areas that have long and complex histories of industrialization or that suffer from regional pollution.

17 Lora Wainwright, 'The Inadequate Life: Rural Industrial Pollution and Lay Epidemiology in China'.

The book also explores the ways in which emotions, economic interests and social structure affect the attribution of responsibility for the impact of pollution on health and the way in which conflicts come to be framed and negotiated between villagers, industry and local governments. The authors note how cancer patients and their families struggle to construct a meaningful explanation for the occurrence of disease and how these emotions, especially in combination with limited scientific understanding and anger toward unresponsive government agencies, can lead them to become fixated on a certain interpretation of the cause of the disease in spite of evidence suggesting more a complex reality. Several of the analyses also explore the internal divisions that arise within villages as a result of the uneven distribution of the costs and benefits of pollution, as, for example, when some villagers own or are employed by polluting industries and others are not.

The study of a waste incineration plant in a village in Guangdong, is particularly interesting. A list of 'cancer victims' was posted on the internet and picked up by media. Journalists visited the village and interviewed angry residents, who complained about noxious smells and dust from the plant. They alleged that dioxin emitted by the incinerator was the cause of rising rates of respiratory system cancer since the plant opened. However, when Chen and his team visited the plant they found it to be clean and well run, with regular testing of emissions; a 'village surveillance' team was allowed to conduct inspections whenever they wished. The CDC reported that they had investigated the reported cancer cases and that the level of cancer was not in fact unusually high.

The researchers realized that conditions in the plant might have recently improved, but they were doubtful that cancers could be attributed to pollution from a plant that had only just opened. Yet, this left them unable to understand the sudden leap in deaths. The answer emerged when Li Qi painstakingly visited each of the households on the list of reported respiratory cancer victims and found that a lot of the information was false, from the age of the deceased to what they died of. Many of those the media claimed had died of lung or other respiratory tract cancers had in fact died of liver cancer or other, completely different diseases unlikely to be related to air pollution. While it was not clear who initiated the list, the media had circulated it without adequate verification, and other people took advantage of the media coverage to oppose the building of another incinerator.

This finding may seem to cast the villagers in a very negative light, but Chen's analysis is more nuanced and again history is important here. The villagers' anger was rooted in their experience with a landfill that was built in 2000, and which caused serious pollution to local water sources.

Villagers' complaints were ignored and they were left deeply suspicious of the authorities. This mistrust later extended to the CDC, which conducted an investigation in the area and did not find anything irregular, but did not attempt to explain the findings to the villagers in language that they could understand or counter the misleading media reports.

These essays thus situate analysis of the impact of pollution on health, and community perceptions and responses to it, within the larger context of the unevenly distributed opportunities and costs that industrialization and urbanization have brought to various parts of China. In some cases, the benefits of industry have accrued mostly to outside investors or landfills and incinerators treating urban people's waste have been located on rural land, with deleterious impacts on the environment and no benefits for the locals. But there are also situations where villagers themselves have polluted the environment with small workshop-style factories or mines. Migration is also part of the story, as some industries draw workers in with them from other counties, while others employ at least some of the local population. The essays show that it is difficult to apply a simple environmental justice analysis to China, because although in some cases there is a clear victim, in many cases victims and beneficiaries overlap. As other studies have found,[18] conflict seems most likely to occur either in cases where locals are deriving little benefit from an existing polluting industry, or when the benefits derived from industry have disappeared, leaving health problems and a degraded environment in their wake.

In media reports and much of the social science literature on environmental pollution in China, local government is either left out of the analysis or cast in a purely negative light, as failing to enforce environmental regulations or respond to villagers' concerns about the impact of pollution on health. The cases in this book also find instances of local government colluding with industry to disguise the effects of pollution, and ignoring villagers' calls for environmental testing or for action against polluting factories. But while not excusing this behaviour, some of these studies also contextualize it with an analysis of the pressures that local government faces to attract investment and generate tax revenue. These include not only performance criteria imposed by higher levels of government, but also the pressure to generate revenue to support public services and policy mandates not covered by central funds. Like villagers, government officials are also not a homogeneous group, and village level cadres in particular face difficult

18 Deng and Yang, 'Pollution and Protest: Environmental Mobilization in Context'; van Rooij, 'The People vs. Pollution: Understanding Citizen Action Against Pollution in China'.

dilemmas as both representatives of the state and members of communities suffering from pollution.

Chen Ajiang's decision not to evade the problem of causality considerably enriches the analysis. As social scientists, we are trained to understand that the aspects of 'reality' we choose to investigate, the categories, methods, and instruments we use, and the kinds of information we privilege or ignore are all shaped by social factors as well as by the more fundamental limitations of our human cognitive abilities. The virtue of Chen's analysis is that he enables us to see the value and limitations of multiple kinds of knowledge and the ways in which circumstances filter and colour what different actors see. As a social scientist who comes himself from a rural background, and who also has some prior training in the natural sciences, Chen is in a particularly good position to give fair treatment to information from different sources, including villagers' lay epidemiologies as well as the limited data available from environmental and health tests. He notes the way in which powerful actors use their control over information to prevent villagers from finding out the risks to which they are exposed, and he understands that their experiential knowledge of the local environment and patterns of disease in the community can enable them to identify changes that may not be captured by formal data collection. But at the same time, he also shows that villagers' own understandings and use of information are loaded in various ways by emotional and social ties as well as by economic interests.

Importantly, while alerting us to the different contexts in which knowledge is generated and used, the authors also make it clear that exploring different answers to the question of causality is not merely an intellectual exercise. Even if we recognize the contingent nature of our understandings of the world, we still have to act, and this entails choosing one explanation over others as a guide to behaviour. The problem of cancer villages puts this tension between uncertain knowledge and the need for action into particularly stark relief and Chen Ajiang does an admirable job in both illuminating and navigating it.

In one sense, it could be argued that there is no need to try to prove causality in individual cases. On a broader level it is painfully clear that the latent impacts of environmental degradation resulting from China's earlier waves of rural industrialization are now manifesting themselves. This is evident from the study of the Huai River Basin and other analysis of large-scale data sets.[19] The case studies in this volume reveal the human suffering behind

19 Yang and Zhuang, *Atlas of the Water Environment*; World Bank, 'The Cost of Pollution in China'.

such impersonal statistics. As the authors point out, the experience of these villages is a warning call to other areas that may be seeking to follow the same development pathway, especially as industry is now relocating from coastal areas to the hinterland and west of China. If the stories in these pages are not to be repeated, regional development policy needs to be proactive in considering the possible environmental and health effects of industrialization and in seeking to reduce the perverse incentives that lead local governments and communities to tolerate pollution. Resources and capacity are limited and cannot be quickly scaled up across the board, but targeted investments could be made in increasing enforcement capacity in areas that are likely to be particularly vulnerable, while local government's fiscal dependence on revenue from industry could be reduced if more services were supported by the central level. Even without such measures, documenting and sharing information about the experiences of early industrializing areas can help to alert other jurisdictions to the long-term consequences of unregulated industrialization and prompt them to assess the environmental and health implications of the development options open to them.

At the same time, though, something must be done in places in which problems have already surfaced, or soon will. And for this we do need to get to grips with the problem of causality in local contexts. This is first because without an understanding of possible pollution sources and pathways of exposure, interventions may be misguided and ineffective. It is important to know whether people are being exposed to toxins from current or legacy pollution and through which environmental media. If current pollution is not a problem and previous contamination affected mostly air or surface water which is now of acceptable quality, providing medical care may be the primary need, but if soil or ground water are contaminated, ecological restoration may need to be considered. In extreme cases, organized migration may be the only way to protect health. But such major interventions are lengthy and expensive undertakings that disrupt livelihoods and communities, and so it is important to target the problem as accurately as possible. Sometimes, understanding environmental media and exposure pathways can inform more immediate measures to reduce health impacts by changing the supply of drinking water or food or changing agricultural practices. But it is still important to identify the main source of exposure or such interventions will not be effective.

Of course, it will not be possible to carry out extensive environmental testing in every case: it is too expensive and there are not enough experts who are trained to do it, especially at the local level. But this does not mean that basic environmental health assessments cannot be carried out, and because many rural communities will be facing these problems in coming years, it will be important to design tools for doing these. Although China has a National Plan for Environment and Health Work which should be implemented at the county level (MEP 2007), not much progress has been made so far, partly because financial and technical resources have not been made available for local environmental and health agencies to conduct this work. As a result, most Centres for Disease Control and Environmental Protection Bureaus at the county level lack the necessary expertise and equipment to conduct environmental health risk assessments.[20] Providing these resources will be essential if localities are to respond better to these problems.

Even then this will not be a simple process because the information generated by risk assessments will be used not just to inform interventions but to determine responsibility and provide the basis for claims for compensation for economic and health damages and clean-up costs. This, of course, is the reason why many local governments are not only unable but also reluctant to carry out these investigations. And it is the reason why, even when they do, villagers often do not accept results that do not substantiate their claims. In fact, although many of these cases are rejected by the courts, in principle, Chinese law takes the position that in cases where polluting events occur, the polluting party must prove that damages to health have not occurred. But as the essays in this volume and other research shows, this law cannot be effective if information about production processes, emissions and environmental quality is not available. At the central level, the Ministry of Environment has recognized the role that the public can play in environmental enforcement, and the recent revision of the Environmental Protection Law has stronger provisions for 'public supervision,' but it is not clear how this will be operationalised at the local level. Yet without better access to trustworthy public information, citizens can only fall back on their own observations; and without proper processes in place to adjudicate competing claims, more and more of these situations are likely to result in conflict.

As this volume shows, in some cases it is possible to establish causality, and in these situations polluting industries should be held responsible for health and other damages. In other cases, the search for a culprit will come up blank because the factories responsible for pollution will have moved or closed, or causality will just be too unclear or complex to attribute responsibility. In these cases, government will have to bear the costs. In

20 Holdaway, 'Environment and Health Research in China'; Su and Duan, 'Zhongguo huanjing yu jiankang gongzuo de xianzhuang, wenti he duice'.

the saddest cases, communities may come to the realization that their own efforts to increase their incomes through small scale mining and industrial activities are the cause of disease. This will be especially distressing in situations where the long-term costs of environmental health impacts far outweigh the short-term benefits that industrialization has brought. It is to be hoped that greater public understanding of these long-term trade-offs will also help to generate support for a more cautious approach to development.

These studies therefore have important implications for policy. In closing, we would like to draw attention to the implications of this book for research. Several of these projects were carried out with support from the Social Science Research Council's (SSRC) China Environment and Health Initiative Collaborative Grants Program, and the team members have been active participants in the China-based Forum on Health, Environment and Development (FORHEAD). Both these programs were established with the goal of promoting interdisciplinary approaches to the study of environmental impacts on health in China with a view to providing a better evidence base for policy and civil society responses.

This book demonstrates very effectively the need for interdisciplinary collaboration. Where many social scientists take the pollution-health relationship as given, or put the question of causality aside, Chen's work shows how crucial it is to gain an understanding of the relevant medical and environmental data in order to unravel what are often complex puzzles; otherwise it is impossible to know how problems might be most effectively addressed or where responsibility lies. In many cases, although a lot of studies on health and environment suggest that causal relationships exist, what they actually show is only a correlation or an association, not causation. This is because health outcomes are shaped not only by physical, chemical and biological factors in the environment, but also by social, economic and cultural factors such as lifestyle and overall levels of health and nutrition. However, identifying such associations nonetheless helps to identify important relationships that require further investigation and when there is an urgent need to act in order to prevent further exposure to health risks, they can also point to possible interventions. Chen's studies acknowledge this complexity. They also demonstrate the importance of social science in surfacing social and cultural factors that can also generate, increase or diminish risks; and in understanding why communities perceive and respond in the ways they do. This knowledge is equally vital to tackling environment and health problems both at the macro level of policy making and in specific communities.

While recognizing the challenges involved, Chen also calls for more research focused on the local level. In his essay on Huangmengying village in Henan, 'Village Perspectives on Understanding the Causes of Disease,'[21] he reflects on the significance of the village as an analytical unit, arguing that it offers a useful middle ground between large-sample statistical data and individual pathology for understanding the environmental sources of disease. This is because the village context makes possible an integrated, in depth study of the interaction between the natural environment and social practices over time. He calls for greater collaboration between medical science and social science researchers in village communities, especially at a time when chronic diseases related to environment and lifestyles are replacing infectious diseases as the major problems in rural China.

This book, now revised to include new material and analysis, presents a highly informative account, not only of the complex world that lies behind the headline 'cancer village', but also of the larger social transformations at work in rural China. It offers no easy answers to the dilemmas they represent, and it does not take refuge in simplistic attributions of blame. As a result, it sometimes makes for very painful reading. But in unpacking the tangled relationships between industrialization, environment and health in these different contexts, it remains a path-breaking contribution to the social science of rural China.

Beijing, October 2016

Coda to the 2020 English Edition

We hope that this book will help a broader audience of non-Chinese speaking readers understand the complex challenges facing rural China and the tangled interactions between development, environment and health, which are often simplified by the media. The introductory chapter, which has been added for this edition, situates the case studies in the context of the profound changes that industrialization and urbanization have brought to the daily lives of China's rural population. As Chen and his colleagues point out, 'cancer villages' may no longer be a hot topic in China, but the tensions between industrial development, intensified agriculture and public health will continue to be salient for the foreseeable future and China's struggles to navigate them also have reference value for other countries facing similar trade-offs.

Beijing, August 2019

21 In this English language edition, this discussion has been included as the final section of Chapter 1, 'Ins and Outs of "Cancer Villages",' to avoid repetition.

Works Cited

Bramall, Chris. *The Industrialisation of Rural China*. Oxford: Oxford University Press, 2007.

Brown, Phil. Toxic Exposures: Contested Illness and the Environmental Health Movement. New York: Columbia University Press, 2007.

Deng, Yanhua, and Guobin Yang. 'Pollution and Protest: Environmental Mobilization in Context.' *The China Quarterly* 214 (2013): 321-336.

Gong Peng, Liang Song, Carlton, Elizabeth J., Jiang, Qingwu, Wu, Jianyong; Wang Lei, Remais, Justin V. 'Urbanization and Health in China.' *The Lancet* 379, no. 9818 (2012): 843-852.

Han, Shi, and Lei Zhang. 'China's Environmental Governance of Rapid Industrialization.' *Environmental Politics* 15, no. 2 (2006): 271-292.

Holdaway, Jennifer. 'Environment and Health Research in China.' *The China Quarterly* 214 (2013): 1-28.

—. 'Environment, Health and Migration: towards a more integrated analysis'. United Nations Research Institute for Sustainable Development Working Paper 2014-3 (March 2014). http://www.unrisd.org/80256B3C005BCCF9/%28httpAuxPages%29/7BCA8ADC1D874482C1257C92004FB1FB/$file/Holdaway.pdf. Accessed 5 March 2020.

Hu, Xiaojiang, Sarah Cook and Miguel Salazar. 'Internal Migration and Health in China.' *The Lancet* 372, no. 9651 (2008): 1717-1719.

Jing, Jun. 'Environmental Protests in Rural China.' In Elizabeth J. Perry and Mark Selden (eds.), *Chinese Society: Change, Conflict, and Resistance*, 143-160. London: Routledge, 2000.

Lora Wainwright, Anna. 'An Anthropology of "Cancer Villages": Villagers' Perspectives and the Politics of Responsibility.' *Journal of Contemporary China*, 19, no. 63 (2010): 77-99.

—. 'The Inadequate Life: Rural Industrial Pollution and Lay Epidemiology in China.' *The China Quarterly* 214 (2013): 302-320.

—, Yiyun Zhang, Yunmei Wu, and Benjamin van Rooij. 'Learning to Live with Pollution: The Making of Environmental Subjects in a Chinese Industrialised Village.' *The China Journal* 68 (2012): 106-124.

MEP (Ministry of Environmental Protection). 'Huaxuepin fangkong shierwu guihua' (12th Five-Year Plan for the Control of Toxic Chemicals). 2013.

MOH (Ministry of Health). 'Guojia huanjing yu jiankang xingdong jihua' (National Action Plan on Environment and Health (2007-2015)'), November 2007. http://www.moh.gov.cn/open/web_edit_file/20071108173502.doc. Accessed 7 December 2012.

—. 2012. '2012 nian woguo weisheng tongji tiyao' (National health statistics report), http://www.moh.gov.cn/publicfiles/business/htmlfiles/mohwsbwstjxxzx/s9092/201206/55044.htm. Accessed 7 December 2012.

Su, Yang and Duan Xiaoli. 'Zhongguo huanjing yu jiankang gongzuo de xianzhuang, wenti he duice' (Current situation, problems and responses in China's environment and health work). In Jennifer Holdaway, Wang Wuyi, Zhang Shiqiu and Ye Jingzhong (eds.), *Huanjing yu jiankang: kuaxueke shijiao* (*Environment and Health: Cross-disciplinary Perspectives*), 72-98. Beijing: Social Science Academies Press, 2010.

Tesh, Sylvia Noble. *Uncertain Hazards: Environmental Activists and Scientific Proof.* Ithaca, NY: Cornell University Press, 2000.

Tilt, Bryan. 'The Political Ecology of Pollution Enforcement in China: A Case from Sichuan's Rural Industrial Sector.' *The China Quarterly* 192 (2007): 915-932.

van de Poel, E., O. O'Donnell, and E. van Doorslaer. 'Is There a Health Penalty of China's Rapid Urbanization?' *Health Economics* 21, no. 4 (2012): 367-385.

van Rooij, Benjamin. 'The People vs. Pollution: Understanding Citizen Action Against Pollution in China.' *Journal of Contemporary China* 19, no. 63 (2010): 55-77.

Wang, M., M. Webber, B. Finlayson, and J. Barnett. 'Rural Industries and Water Pollution in China.' *Journal of Environmental Management* 86, no. 4 (2008): 648-659.

Wang, Wuyi, Linsheng Yan, Hairong Li, and Yonghua Li. 'Zhongguo huanjing bianhua fengxian guanli duice' (Management responses to health risks arising from environmental change in China). In Jennifer Holdaway, Wang Wuyi, Zhang Shiqiu and Ye Jingzhong (eds.), *Huanjing yu jiankang: kuaxueke shjiao* (*Environment and Health: Cross-disciplinary Perspectives*), 25-39. Beijing: Social Science Academies Press, 2010.

World Bank. *Toward a Healthy and Harmonious Life in China: Stemming the Rising Tide of Non-Communicable Diseases.* Washington, DC: World Bank, 2011.

World Bank and State Environmental Protection Agency. 'The Cost of Pollution in China: Economic Estimates of Physical Damages'. Washington, DC: World Bank, 2007. Available at www.worldbank.org/eapenvironment.

Yang, Gonghuan and Zhuang Dafang, eds. *Atlas of the Water Environment and Digestive Cancer Mortality in the Huai River Basin.* Dordrecht: Springer, 2013.

Zhou, Maigeng, Haidong Wang, Jun Zhu, Wanqing Chen, Linhong Wang, Shiwei Lu, et al. 'Cause-Specific Mortality for 240 Cases in China During 1990-2013; A Systematic Subnational Analysis for the Global Burden of Disease Study 2013.' *The Lancet* 387, no. 10015 (2016): 251-272.

1 Retrospective Thoughts on the 'Cancer Village' Phenomenon

Chen Ajiang

From a long-term perspective, the phenomenon of 'cancer villages' may perhaps turn out to be a fleeting one in the history of environment and health in China. The 2004 CCTV (China Central Television) programme 'Rivers and Villages' made the issue a hot topic and drew public attention to other mentions of 'cancer villages' that had been circulating over the previous few years. Five years later, in 2009, the National Key Laboratory for Environmental Chemistry and Ecological Toxicology of the Chinese Academy of Sciences organized the 29th International Symposium on Halogenated Persistent Organic Pollutants – Dioxin in Beijing. After the Symposium, CCTV broadcast a programme titled 'International Symposium on Dioxin Finds Waste Disposal Encountering Problems, and Dioxin is Troubling China'. Waste incinerators became a focus of contention and 'cancer villages' became a hot topic again. In 2013, the Ministry of Environmental Protection's '12th Five Year Plan for the Prevention and Control of Chemical Environmental Risks' formally acknowledged the problem of 'cancer villages', bringing the question of whether they existed, which had persisted for about ten years, to a sort of conclusion.[1]

If we take a ten-year time frame, we can more or less distill the social logic behind the emergence of the topic of 'cancer villages'. After reform and opening up began in 1978, China's economy developed rapidly and with this came progressive worsening of environmental pollution. This had serious negative impacts on the ecological environment, which in turn affected the economy and society. China is a huge country, with great regional diversity, so it is not surprising that some places would have a higher incidence of

1 The years 2012-2015 could be seen as an important turning point for China's governance of environmental pollution. The government's acknowledgment of the existence of 'cancer villages', shows that environmental pollution problems in China started to get serious attention.

Chen Ajiang, Cheng Pengli, and Luo Yajuan, *Chinese "Cancer Villages": Rural Development, Environmental Change and Public Health*. Taylor & Francis Group 2020
DOI: 10.5117/9789089647221_CH01

cancer than others. At the same time, there was an imbalance of power in the ability of different social actors to deal with pollution. Local governments, under pressure to increase GDP (Gross Domestic Product), and driven by their own interests, allied themselves with enterprises, making it hard for ordinary people affected by pollution to fight back. What is interesting, is that this was going on precisely at the time when new media were taking off. As commercial enterprises, the media have to find something to make a noise about to make their presence felt, and members of the public who were helpless in the face of the impacts of pollution made use of the new media as a tool of resistance. They not only provided a platform for social networking, but also an outlet for voices that had been repressed. Sometimes, they even acted as a platform for 'distorting the tone' (*bianyin*) of the debate. The news in these media was very diverse and some people accepted and spread information that was in their own interests, in some cases even deliberately creating false rumours. As a result, in some regions, the public's lack of trust in enterprises and local government reached an unbelievable level.

In January 2018, when the authors of this volume were planning the English edition of this book, we decided that it would be helpful to include an introductory chapter. This was partly to provide background for non-Chinese readers, but also because most of the research discussed in the book was conducted around 2010, when cancer villages were a hot topic. Now that moment has passed, we can reflect on this issue in a more measured way.

In the sections that follow, I give a short introduction to the process of conducting the 'cancer village' research reported in this book, as well as to some of the key themes that cut across the specific case studies. These include questions relating to the relationship between pollution and the risk of cancer, the social construction of 'cancer villages' and the 'post-cancer village' era.

Origins of the Study and the Research Process

It was probably around 1970 when people in the little village where I grew up started using chemical pesticides. At that time, every household had to store a week's supply of water, because for a week after the chemicals were sprayed, villagers were not allowed to use water from the river for drinking or bathing. Pesticides were sprayed only on the paddy fields, so it might seem there was nothing to worry about, but in fact it was hard to ensure that the water from the paddies did not leach through the irrigation ditches into the river. After pesticides were used, all the fish, shrimp and other creatures that lived in the paddies and ditches were wiped out and dense piles of dead fish floated on the surface of the water. Before this, the things villagers most feared were droughts and floods, poisonous snakes, wild beasts, and ghosts and evil spirits. This was the first time they had encountered industrial products that threatened the environment on which they depended for their basic survival.

As time passed, the horror of pesticides faded from the minds of the villagers, but with the appearance of Township and Village Enterprises after reform and opening up in the late 1970s, the impact of industrial pollution on water bodies worsened. After 1990, the water in rural lakes and rivers went from occasionally having a strange smell to being totally unfit for human consumption. Villagers were constantly asking themselves, 'Can we actually drink this water? Can we eat these fish? Can we use river water to rinse our rice?' Rural people became progressively more helpless in the face of pollution and increasingly worried about the risks it presented to their health.

It was in 2004 that I first had the opportunity to undertake a systematic study of pollution with the project, 'Water Pollution in the Tai Basin: a Sociological Perspective', which was funded by the Jiangsu Province Social Science Foundation.[2] Water pollution in Lake Tai had been getting worse since the turn of the century, and when conducting research in southern Jiangsu and northern Zhejiang I found that local officials and ordinary people were all very worried about the possible harm that pollution could cause to their health. I heard people mention 'cancer villages' and say things like 'Not a single boy in the village passed the health test for the military.' This had to some extent be an exaggeration, but at the time I had no way of proving it either way, and I had no means of doing in depth research on this topic. However, such comments did alert me to the issue of the impact of pollution on health.

In 2007, my team and I began a new project funded by the National Social Science Council, titled 'Harmony between People and Water', which involved fieldwork in villages around Lake Tai and in the Huai River Basin. In this project, I constructed two ideal types, 'lack of harmony between people and water' and 'harmony between people and water'. In communities in which there was a lack of harmony between people and water, water pollution was

2 The final result of this project was the book *Secondary Anxiety: A Sociological Perspective of Water Pollution in the Tai Basin,* published by China Social Science Press in 2009.

affecting people's health and causing disease. Pollution was also affecting economic development and causing poverty. Once these various negative impacts of pollution reached a certain level, people affected by it would begin to 'flee' the area. This out-migration would further fragment the community, leading to secondary social problems including inequality. Dongjing Village in Funing County in Jiangsu, Xiqiao Village in Jiaxing City of Zhejiang, and Huangmengying Village in Shenqiu County in Henan were all cases of a 'lack of harmony between people and water'.

In 2010 we began another project, 'Perceptions and Responses to Environmental Health Risks: based on fieldwork in several cancer villages', with funding from the Social Science Research Council (USA). Apart from the three cases studied in the previous research, we chose additional sites in Jiannan Village in Yiyang County in Jiangxi, Shangba Village and Liangqiao Village in Wengyuan County in Guangdong, along with a few other villages. In designing the study, we reviewed the research approaches of other scholars, and attempted to put to one side questions that related to natural science, especially those questions on which natural scientists had not yet reached a consensus. Our idea was to put contested questions of natural science aside and do 'pure social science' research. But as soon as we went out into the field in December 2010, we realized that it was actually impossible to set aside or ignore questions about the severity of pollution, the incidence of disease and the relationship between them. The people we were researching were constantly raising questions and arguing about 'pollution' and 'cancer' and their relationship, and it was impossible for us to avoid the issue. Therefore, both in the process of the research and in the presentation of the results, we did not shy away from natural science problems. Instead, we did everything we could to understand the state of existing natural science knowledge on these questions and to push it forward where we could. We took the position that we should make full use of the findings of natural science, including published papers and test reports by relevant government agencies, and do our best to clarify on a scientific level what the situation was in our study sites in terms of the variable relationship between pollution and disease. When this relationship was not clear, we tried to arrive at the best possible understanding and minimize the likelihood of bias by specifying the range of possible uncertainty with regard to a few key variables.[3] We used a variety of research methods and

3 For example, in Jiannan, we analysed every pollutant that villagers mentioned as a cause of cancer. With regard to test results that showed well water to be below the standard, we looked up relevant literature and consulted scientific experts. With regard to the suspicion

conducted field investigations to gather all kinds of information, which we then analysed in an integrated way. We attempted to take into consideration the opinions of all the relevant groups of people and understand the positions and attitudes of ordinary people as well as government agencies, emphasizing the importance of understanding individuals, society and culture.

With the publication in 2013 of the Chinese volume *Cancer Village Research*, our field research on cancer villages reached a natural pause. I hoped that after this I would be able to make some kind of breakthrough in terms of the methods used in 'cancer village' research. I read widely in the medical and epidemiological literatures, and in environmental history, and this gave me the opportunity to step outside the limitations of methods that were based primarily on an experiential approach. The article 'Environmental Change and Health Risks' (Chapter 4 of this volume) was my attempt to apply new methods to research on environment and health. Taking a historical perspective, and particularly one that integrated environmental and social history, I found that some current problems probably had a historical connection. Cancer is a 'slow' disease, and the factors that influence it may have been present for a long time. The disease is also the result of many factors, and so a historical understanding of health risks is helpful in gaining a comprehensive understanding of cancer. At the same time, it is also useful in terms of understanding the conflicts and struggles between the villagers and the local government. This method led me to rethink some of the environmental risks these 'cancer villages' had faced. 'The Ins and Outs of a "Cancer Village"' (Chapter 2 of this volume), and the revised English version of 'A *Subei* "Cancer Village"' (Chapter 3 of this volume), demonstrate the characteristics of this approach. Another member of our team is currently developing this line of research further.

The relationship between the chapters in this book is roughly as follows. 'The Ins and Outs of a "Cancer Village"' discusses the problem of cancer in Huangmengying Village from the perspective of exploring the causes of disease, and the value of community-based research methods in understanding

that pollution from the phosphate fertilizer plant could be causing cancer, we made an effort to understand the details of the plant's production technology and made a judgment based on my own previous scientific knowledge. We visited the County Centre for Disease Control, to understand the result of investigations that they had already conducted, and we went to the Environmental Protection Bureau to obtain more expert background information. (See Chapter 9, 'Perceptions and Responses to Pollution', and Chapter 6, 'Coexistence of Poverty and Cancer').

the multiple factors that contribute to disease. From the point of view of research methodology, it is a retrospective consideration and summary of the empirical methods I used in studying 'cancer villages', and also an implied critique of some of the published mainstream epidemiological research on cancer villages in China.

'Environmental Change and Health Risks' (Chapter 4) presents the case of Lianshui County in Northern Jiangsu (*Subei*). This place is rather similar in terms of its natural history and social characteristics to Dongjing Village, which is discussed in 'A *Subei* "Cancer Village"', and reading the two pieces side by side may make it possible for readers to appreciate the complexity of the phenomenon of 'cancer villages'.

'Villagers' Perceptions of and Responses to the Relationship between Cancer and Pollution' (Chapter 9) is a summary and distillation of general observations from three specific cases which are discussed separately in 'A Prosperous "Cancer Village"', 'Coexistence of Poverty and Cancer', and 'Problematization and De-Stigmatization' (Chapters 5, 6, and 7). While trying to understand the scientific relationship between pollution and cancer, we also explore the social science dimension of the phenomenon of 'cancer villages'. Chapter 8, 'Behind the High Incidence of Lung Cancer', shows the readers another aspect of the phenomenon – the social construction of 'cancer villages' – and, like all the studies, demonstrates the complexity of the issue.

The remainder of this introduction provides some background for readers by sketching a broad picture of the economic and social changes that form the backdrop to the emergence of 'cancer villages' in China as both a scientific and a social problem.

Rapid Industrialization, Pollution and Carcinogenic Risks

In the 70 years since the establishment of New China in 1949, China's economy has undergone massive expansion. The country's GDP in 1980 was 6.8 times what it was in 1952, and by 2015 it was 150 times its 1980 level and 1015 times what it had been in 1952. The gross value of industrial production increased even more rapidly.[4] In 1980 it was almost 17 times its 1952 level, and in 2015 it was 117 times the 1980 level and 1974 times the 1952 level (see Table 1.1). The curve in Graph 1.1 shows the trend graphically.

4 China started to plan its economic development in 1952. Before that date there are no complete economic statistics.

Table 1.1 China's economy from 1952 to 2015

Year	*GDP (billion yuan)*	*Gross Value of Industrial Output (billion yuan)*	*Per Capita GDP (yuan)*
1952	67.9	12.0	119
1955	91.0	19.1	150
1957	106.8	27.1	168
1960	145.7	56.8	218
1965	171.6	54.7	240
1970	225.3	82.8	275
1975	299.7	125.5	327
1978	367.9	162.2	385
1980	458.8	201.5	468
1985	909.9	347.8	866
1990	1887.3	690.5	1,663
1995	6140.0	2502.4	5,091
2000	10028.0	4026.0	7,942
2005	18731.9	7796.1	14,368
2010	41303.0	16512.6	30,876
2015	68905.2	23650.6	50,251

Source: National Bureau of Statistics http://stats.gov.cn.

Graph 1.1 China's economy 1952-2015

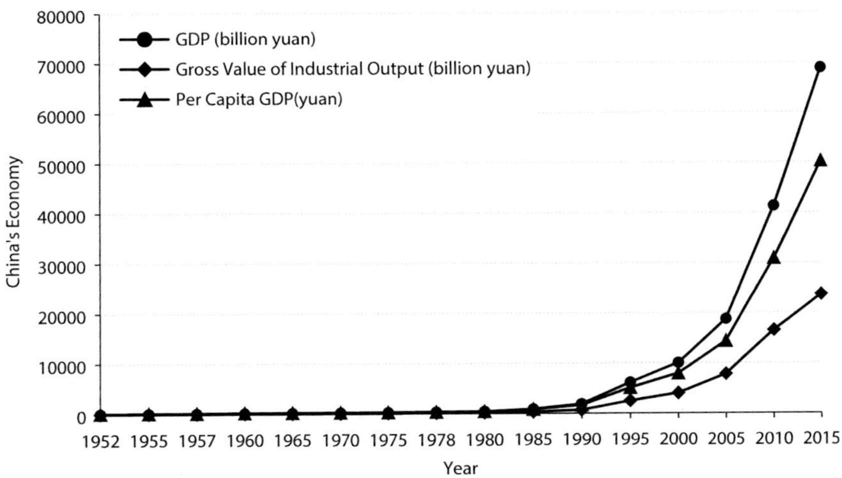

This massive increase in industrial production had a heavy environmental price. The chain begins with the extraction of resources. Industrial production requires large quantities of minerals and it depends on coal and oil to meet its needs for energy. Steel and ferrous metal production requires the mining of metal ores, and cement and other construction materials require mining of stone of various kinds. Extractive processes damage the environment; open-face mining destroys surface vegetation, and the waste water and tailings produced by mining also affect the environment. The refining and processing of minerals have further environmental impacts.

The manufacturing of industrial products generates waste water, gases, and solid waste that directly affect the environment. China's manufacturing industry developed very rapidly in the 30 years after reform and opening up; but the development of technology and management systems for pollution control did not keep pace. In the areas that led this growth in manufacturing, namely the Pearl River and Yangtse River Deltas, environmental pollution was truly alarming. The early commune and production brigade enterprises and later the Township Enterprises, as well as the small privately-owned enterprises that appeared after reform, were mostly at the county level or below. There was no rational planning for these industries and no consideration of environmental impacts in their siting; small industries were scattered throughout the countryside, their production technologies were basic, and they generally had no pollution mitigation equipment. A situation developed in which 'every village has a factory and every house is belching smoke'. After the systemic changes that took place in the late 1990s, including privatization of collective industries and the development of private sector businesses, enterprise owners also had very different values both from those of traditional China and from those that were promoted during the socialist era. They were focused solely on profit and most of them had no sense of social responsibility. This manifested itself in their attitude towards environmental protection. Some enterprise owners, in pursuit of profit, would go to any lengths to reduce their production costs and released pollutants directly into water bodies.

It is important to note that China's economic development has been very uneven. During certain periods, the rate of economic growth in some parts of the country, including the Pearl and Yangtse River Deltas, was several times higher than the national average. In the absence of an effective system of environmental protection, it was inevitable that rapid industrialization in these areas would cause environmental pollution. Places like the Tai Basin, which had formerly been known as the 'land of fish and rice', became places where water was scarce due to poor water quality.

These high levels of pollution inevitably led to an increase in environmental and occupational diseases. A typical case of mining-related disease is the high incidence of lung cancer among mine workers in Gejiu County in

Yunnan. As early as the 1980s, research showed that in the previous twenty or so years, 1,200 people in the Yunxi Company and the municipally owned local mine had died of lung cancer, of whom 90% worked in the mine pits. The mortality rate from lung cancer among miners was almost 35 times that of ordinary people and it increased with the number of years working in the mine. Researchers thought that radon daughters were the main cause of the high lung cancer incidence amongst Gejiu miners, with arsenic in mineral dust, including arsenic pollution in the surface environment, as important compounding factors.[5]

The case of Shangba village in Guangdong included in this book (see Chapters 7 and 9), is located in the multiple metal mining area of Dabaoshan, which is under the jurisdiction of Shaoguan City. Excavation began in 1958 and production started in 1966. In 1995, during the modernization of the enterprise system, the mine was converted to a limited liability company. Dabaoshan is a large, multi-metal mine containing iron, copper, sulphonium, lead, zinc, molybdenum, tungsten and other minerals. According to information on the company's website in 2015, in that year it had an annual production volume of 600,000 tons of iron ore, 3600 tons of cuprous metals and 100,000 tons of sulphate. It also processed 10,000 tons of copper.[6] The first report of a 'cancer village' problem in the area was in 2001,[7] and starting in 2004, a lot of articles appeared about environmental pollution in Dabaoshan.[8] Recent scientific research has confirmed the relationship between pollution from the mine and high rates of cancer. For example, the research of Guan Yanyan and her colleagues demonstrated that blood levels of copper and zinc were higher among population groups who lived downstream from the mine, who had higher levels of exposure to pollution, than among population groups living upstream, and copper and zinc levels were shown to correlate with the incidence of cancer.[9]

5 Sun et al., 'Etiological Analysis of Yunnan Gejiu Tin Miners' Lung Cancer', pp. 1-8, 12.

6 For specific production data, see Guangdong Dabaoshan Mining limited company's official website: http://www.gddbs.com.cn/gsgk.htm. Accessed December 31, 2015.

7 Zhu, 'Waste Mining Water Harmed the Whole Village'.

8 Searching 'Dabaoshan' in the database produced around 300 papers, amongst which studies on environmental pollution were the main source of papers after 2004. The top ten papers ranked by the number of times cited (from 221 times to 54 times) are all about research on environmental pollution. The range of research includes water and agricultural soil pollution from multiple-mineral mines, the health and social influence of environmental pollution, and corresponding strategic responses.

9 Guan et al., 'Using ICP-AES to Test the Level of Chemical Element Cu, Zn, Fe and Mg Contained in the Whole Blood of People who are Under High Exposure to Minerals'.

There is a clear relationship between levels of industrialization and lung cancer. As the '1973-1975 Survey of Mortality from Malignant Tumours' shows, the places in which lung cancer was among the top five causes of death were Shanghai, Tianjin, Beijing, Liaoning and Jilin.[10] In the 1970s and before, these provincial level cities and the North-eastern region (where Liaoning and Jilin are located) had the highest levels of industrialization. The 1990-1992 'Survey of Mortality from Malignant Tumours' shows that 'along with socioeconomic development and medical advances, the mortality rate from some cancers fell; while some cancers became more prevalent due to pollution of the environment, widespread use of chemicals, and an increase in the number of people smoking cigarettes. Lung cancer was one of those that increased. Morbidity and mortality from lung cancer currently shows an annual increase, and especially in major cities, it has already become the leading cause of death from cancer'.[11] In 2013, the mortality rate from lung cancer ranked first among deaths from malignant tumours, with an incidence rate of 53.37/100,000 people and a mortality rate of 45.57/100,000 people. This is almost ten times the mortality rate from lung cancer forty years ago.[12]

Some of the cases in this book are ones in which rural industry has polluted water sources. For example, Xiqiao Village is in the central region of the Yangtze River Delta that industrialized very rapidly, and villagers suspected that the high incidence of cancer was related to factories that were polluting the river water (see Chapters 5 and 9 of this volume). In 'A *Subei* "Cancer Village"' (Chapter 3), the villagers also pointed to factories at the edge of the village as the cause of the high cancer rate. The high incidence of cancer in the Huai River Basin has also been attributed to industrial pollution of the river, but the actual situation may be more complicated than that (see Chapter 2 of this volume).

Agricultural Development, Nitrogenous Substances and Cancer risks

Although when we think about pollution and health, it is natural to think of industry as the main culprit, agriculture is also a major source of pollution.

10 'National Office of Tumour Treatment and Prevention Research', *National Survey of Mortality from Malignant Tumours 1990-1992*, p. 154.

11 National Office of Tumour Treatment and Prevention Research.

12 People.cn. 'Our Country's Mortality Rate from Lung Cancer Increased Almost Ten Times Within 40-Year Period'.

This is partly due to the use of pesticides, which are designed to be toxic. A less well-understood question, which has become the focus of research in recent years, is the relationship between various nitrogenous substances and cancer.

Nitrogen compounds including nitrosamines and *N*-nitrosamide are broadly carcinogenic and cause cancers in all the organs and connective tissues of all animals. Nitrosamines which have a symmetrical chemical structure cause oesophagus cancer, and those with an asymmetrical structure cause liver cancer, while *N*-nitrosamide causes stomach, oesophagus and other cancers. Much of the research on the relationship between nitrogenous substances and cancer centres on the 'Nitrogen Cycle Hypothesis', which was put forward by Xu Zhixiang in 2003 and has since received some empirical support in medical research and from the implementation of water improvement projects.[13] The hypothesis focuses on the formation of nitrosamines and explains how their precursor elements are generated from the environment and enter the human body.

The basic points of the hypothesis are as follows. Nitrates and nitrites, along with amines and amides in waste water, enter the body through drinking water, and combine to form nitrogenous compounds. Nitrates and nitrites can transform into each other in the environment. Nitrogenous substances include, primarily, naturally occurring microbes in the waste products and remains of animals and plants and rural household waste; chemically synthesized nitrogenous fertilizer; and waste water from food and paper industries. In the nitrogen cycle, protein eventually degrades into nitrates, and during the process it produces amines and amides. Chemically synthesized nitrogenous fertilizer also breaks down in the natural environment to create nitrates and nitrites. The urea in nitrogen fertilizer is basically an amide precursor.

Over the last half century or so, there has been a big increase in the quantity of nitrogenous substances in rural areas, and their spatial distribution has changed in ways that increase the possibility of nitrogenous compounds, including nitrosamines and nitrosamides, getting into the drinking water, increasing the risk of cancer.

First, the total amount of nitrogen in the environment has increased. The use of chemically synthesized nitrogen fertilizers to increase agricultural output has clearly increased. The government began to promote the use of chemical fertilizers starting in the 1960s, but initially the total quantity

13 See Xu (ed.), *Fertilizer, Waste Water and Oesophagus Cancer*; Xu et al., 'The Nitrogen Cycle Hypothesis and the Cause of Oesophagus Cancer, Stomach Cancer And Liver Cancer', pp. 61-74.

Graph 1.2 Chemical fertilizer (in converted tons)[16] and net amount of nitrogen fertilizer used in agriculture in China, 1980-2015

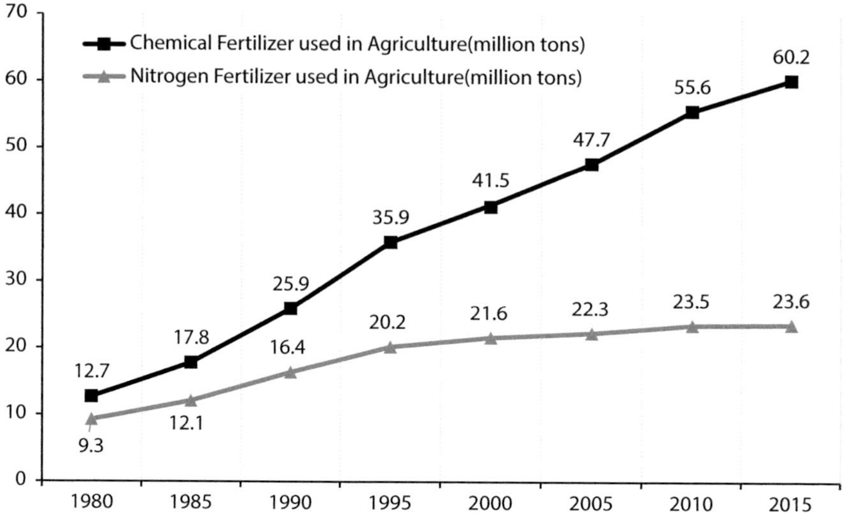

Source: National Bureau of Statistics http://stats.gov.cn

used was small. In 1949, the total amount of chemical fertilizer used in China was only 6,000 tons;[14] by 1965 it had increased to 1.9 million tons.[15] After reform and opening up, the use of fertilizer continued to increase. In 1980, around 12.7 million tons was used in agriculture, of which nitrogen fertilizer accounted for more than 9.2 million tons; by 2015 this had increased to more than 60 million tons, or 4.7 times the amount used in 1980, and 31 times that used in 1965. See Graph 1.2.

Second, the total amount of nitrogen fertilizer lost to the environment in China is quite substantial. The 2010 'Report on China's First Pollution Census' showed that in 2007 almost 1.6 million tons of nitrogen fertilizer were lost, or 59.8% of all nitrogen found in agricultural pollution. Of this, 320,100 tons were washed away through surface channels, and 207,400

14 China Industry Information Net, 'General Analysis of China's Fertilizer Industry's Financial Operation'.

15 National Bureau of Statistics, *China Statistical Yearbook 1985*, pp. 281.

16 Converted tons measures the amount of nitrogen in nitrogenous fertilizer, and for phosphate fertilizer the potassium and phosphate contents are measured separately. Statistics use these converted measures, not total volume of fertilizer used.

tons leached into the ground.[17] Furthermore, the Nanjing Soil Institute of the Chinese Academy of Sciences collected materials from 782 field experiments conducted across the country, and found that the utilization rate of nitrogen fertilizer in China was only 33.3%.[18] The unused fertilizer mostly ends up being washed into rivers and lakes, or underground, polluting water bodies.

Another problem is that water engineering projects have changed the spatial distribution of nitrogen, making it more likely to get into drinking water bodies. Starting in the late 1950s, China embarked upon numerous water conservancy projects, ranging in scale from massive projects to control major rivers, to the construction of small irrigation ditches in fields. These modernization projects transformed water courses, and in the process also changed the spatial and geographical distribution of nitrogenous substances. For example, in the Huai River Basin, efforts to control the river that began in the 1950s fundamentally changed the temporal and spatial distribution of water flows in the area. For Huangmengying Village (see Chapter 2 of this volume), this meant that unused nitrogen from industries located upstream, including food and paper factories that processed grain and wood pulp, was released as waste and trapped in the river courses above the Huaidian Sluicegate as if in a reservoir. The polluted water then flowed through irrigation ditches into the ponds of the village, and from there through ditches into agricultural fields. Polluted water also leached into shallow wells, which were the source of villagers' drinking water, increasing risks to health.

Northern Jiangsu was affected by the changed course of the Yellow River, which resulted in it becoming an impoverished region. To address this situation, after the 1950s, the government embarked on a large-scale water engineering project. This enabled the formerly dry land to be put under rice paddy cultivation, greatly increasing grain production, and it also solved the problems of flooding and drought. But in bringing the river under control and changing the crop structure, the project also brought changes in the living environment of the villagers. Their drinking water ponds were now connected to the paddy fields, and the nitrogen-rich water could flow into these ponds, inadvertently causing a health risk (see Chapter 4 of this volume).

17 Ministry of Environmental Protection, State Statistical Bureau, and Ministry of Agriculture, *Report on China's First Pollution Census*.
18 Pei, 'Experts Talking About how to Improve the Utilization Rate of Fertilizers', pp. 10-11, 14.

There are many similar situations where village water courses were changed in order to benefit agricultural production. Prior to 1960, the fields in Yangzhong County in Jiangsu were irrigated naturally by water from the Yangtze. But after 1960 they were mostly irrigated with electric pumps, and as a result of this, water in the ditches became stagnant or semi-stagnant. Villagers were drinking stagnant water for more than nine months of the year, and this change increased the risk of cancer: The mortality rate of oesophagus cancer in Yangzhong County reached 101.68/100,000 people, and in Gaoyue Commune it reached a high of 144/100,000 people.[19]

In addition to this, industrial production that uses agricultural raw materials has changed the temporal and spatial distribution of nitrogen. The emissions of waste water from the paper and food industries mentioned above are an example. Because these industries employ relatively simple technologies and can also take full advantage of local raw materials, from the 1980s onwards many rural areas encouraged their development. For example, the pollution in the Huai River Basin in the 1990s was closely related to waste from the paper, food and leather processing industries in the area.[20] Because these industries used harmless agricultural products as raw materials, or grains that could be eaten directly, villagers did not realize the dangers associated with this pollution. In the case of Huangmengying Village, it was outsiders who drew attention to the cause of the problem (see Chapter 1) and in Jiannan Village, the villagers did not recognize the risk from waste water emitted in the process of making MSG (Monosodium Glutamate), and instead pointed to the glass factory and the phosphate fertilizer factory as the main sources of pollution (see Chapters 6 and 9).

19 Xu (ed.), *Fertilizer, Waste Water and Oesophagus Cancer*, pp. 106-107.

20 See Yang and Zhuang (eds.), *Atlas of the Huai River Basin Water Environment: Digestive Cancer Mortality*. The research by Yang Gonghuan and Zhuang Dafang et al. shows the correlation between pollution in the Huai River Basin and cancer incidence: 'Using monitoring data from every water section, we mapped the distribution of the main stream, primary tributary, secondary tributary and lakes over different time periods and regions, and found clear characteristics of distribution and change in pollution in the Huai River Basin. Meanwhile, when comparing changing trends in the mortality structure of the population living in the Huai River Basin over a 30-year period, we found that the areas that had the most severe and lasting pollution were also the areas that had the most rapid increase in the mortality rate from oesophagus cancer; the increase is several times higher than the national average mortality rate for that cancer. The most important finding of this mapping is spatial analysis that shows the spatial distribution of areas that recently have high oesophagus cancer incidence rate is very similar to the distribution of the most severely polluted areas'.

Graph 1.3 China's urbanization rate, 1980-2015

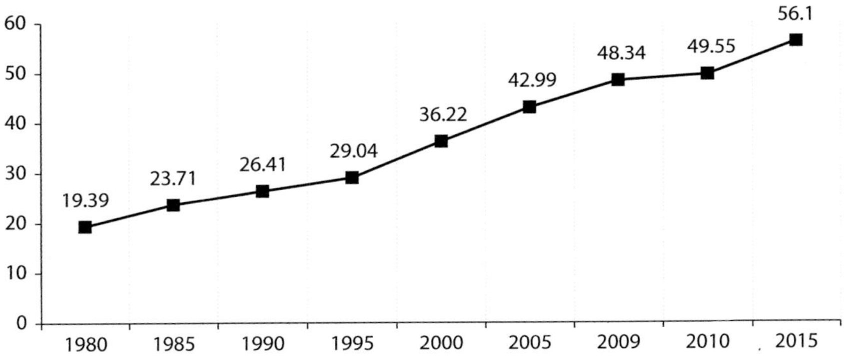

Source: National Bureau of Statistics website http://data.stats.gov.cn/

Urbanization, Household Waste, and the Social Construction of Cancer Risks

During the decade in which 'cancer villages' were a hot topic, the phenomenon of power-generating waste incinerators (often referred to as waste-energy plants) causing cancer was also a big part of the story. Waste incinerators were, in turn, the result of the large volumes of household waste generated as the result of urbanization and higher living standards. Prior to the 1980s, the vast majority of the Chinese population lived in villages. As Graph 1.3 below shows, in 1980 the urban population was not even 20% of the total. Along with rapid economic growth, urbanization sped up, and by 2010 the urbanization rate had already reached 50% (49.95%). In 2015, the urban population accounted for 56.1% of the total population. See Graph 1.3.

As the urban population increased, so did the volume of urban waste, and the problem of dealing with it gradually became very conspicuous. As Table 1.2 shows, in 2015, the volume of waste collected from urban areas reached more than 191 million tons, or 4.3 times as much as in 1985. As people's living standards improved, the composition of waste also became more complex. Previous methods of returning it to the soil or simply burying it were no longer appropriate. On top of that, population density in the more developed areas of Eastern China was very high, and land was scarce and expensive. For these reasons, waste incinerators were quickly adopted as a new means of disposing of waste. The proportion of household waste that was incinerated rose from 5% in 2005 to 10% in 2008, doubling in three years. Four years later, in 2012, it had doubled again to more than 20%. In 2015, around a third of all household waste was being incinerated.

Table 1.2 Household waste: total collected, number of waste incinerators and incineration rate, 2003-2015[21]

Year	Net volume of collected waste (million tons)	Household waste incinerators	Volume of waste incinerated (million tons)	Incinerated waste as percentage of total collected waste (%)
2003	148.6	47	3.4	2.29
2004	155.1	54	4.5	2.90
2005	155.8	67	7.9	5.08
2006	148.4	69	11.4	7.66
2007	152.2	66	14.4	9.43
2008	154.4	74	15.7	10.17
2009	157.4	93	20.2	12.85
2010	158.0	104	23.2	14.65
2011	164.0	109	26.0	15.85
2012	170.8	138	35.8	20.98
2013	172.4	166	46.3	26.88
2014	178.6	188	53.3	29.84
2015	191.4	220	61.8	32.26

Source: National Bureau of Statistics website http://data.stats.gov.cn/

The speed of development in the eastern coastal provinces was even faster. For example, in 2005 the proportion of household waste incinerated in Jiangsu was only 5%, but ten years later, in 2015, it had increased to 72%. Incineration had already become the primary means of household waste disposal. The development of incineration was also very rapid in Fujian and Zhejiang, see Table 1.3.

Before 2009, waste incineration and waste-energy plants were new in China, and most people were not familiar with the technology they used or with their possible environmental impacts. However, shortly after the 29th International Symposium on Dioxin, in September 2009, the CCTV show 'Half an Hour on Economics' informed the audience about the toxicity of dioxins and their potential risk to people's health.[22] The programme explained the dangers that waste incinerators could cause and introduced

21 Data on household waste mainly refers to urban household waste. But in recent years, some regions collected and disposed of urban and rural household waste together, so the data includes rural household waste. Most of the statistics given here are for urban waste but for this reason we have not specified.

22 Dioxins are Persistent Organic Pollutants (POPs) that are produced by a number of industrial processes and by the burning of waste at insufficiently high temperatures. They are highly toxic and can cause reproductive and hormonal problems as well as cancer. Dioxins and their

Table 1.3 Incineration rates over various years as a percentage of all waste collected in three provinces

Region	*Year*	*2005*	*2010*	*2015*
Jiangsu	Volume of household waste collected (million tons)	8.3	10.2	14.6
	Volume of waste incinerated (million tons)	0.4	4.6	10.5
	Incinerated waste as percentage of total collected (%)	5.04	45.10	72.03
Zhejiang	Volume of household waste collected (million tons)	7.6	9.6	13.3
	Volume of waste incinerated (million tons)	2.0	4.4	7.7
	Incinerated waste as percentage of total collected (%)	15.89	45.65	58.03
Fujian	Volume of household waste collected (million tons)	7.7	4.2	6.1
	Volume of waste incinerated (million tons)	0.3	1.4	3.7
	Incinerated waste as percentage of total collected (%)	9.54	34.02	60.19

the different opinions of supporters and opponents of incineration. It also explained the current situation regarding waste disposal in China and trends in the development of waste incinerators. For example, it reported that Beijing planned to rapidly expand its waste incineration capacity to treat 40% of all waste by 2015, and that Zhejiang had already built or was in the process of building 30 waste incinerators, while Guangdong had almost 20, and Jiangsu 15. Meanwhile, the situation in the United States was exactly the opposite: the programme noted that 'Statistics show that in America, the number of waste incinerators has fallen from a high of 171 to the current 102, and not a single new incinerator has been built since 1995'.[23]

2009 was a year of great symbolic significance. In 2009, China's urbanization rate reached 48.34%, which means that almost half the population lived in urban areas. In that year, the total amount of urban waste collected was more than 157 million tons. The rapid rate of urbanization and the lag in the construction of waste treatment facilities had resulted in 'cities being encircled by waste'. By that time, there were 93 waste incinerators and more than 20 million tons of waste were being incinerated, accounting

Effects on Human Health (WHO, https://www.who.int/en/news-room/fact-sheets/detail/dioxins-and-their-effects-on-human-health).

23 CCTV Programme, 'International Symposium on Dioxin Finds Waste Disposal Encountering Problems, and Dioxin is Troubling China'.

for almost 13% of the total. The eastern more developed provinces were far ahead of the average, with almost one third of waste in Jiangsu and Zhejiang being incinerated. Public grievances against enterprises and local government as the result of land grabs, demolitions, environmental pollution and other social problems related to incinerators had been mounting for a long time.[24] In this context, the TV programme about the Dioxin conference in September 2009 served as a kind of wake up call, and on a certain level even had the effect of encouraging the public to oppose waste incineration.[25]

Later analysis shows that in anti-incinerator incidents, conflicts of interest and local pollution were the basis of conflict, but dioxin also acted as a fuse.[26] Conflicts over the acquisition of land, demolitions, forced migrations and resettlement were major underlying causes. The process of urban expansion swallowed up large amounts of rural land and this not only affected rural livelihoods, but also increased peasants' feelings of inequality and exploitation. This was especially true when rural land that was commandeered from peasants for only a few ten thousand RMB was later sold off for a few hundred thousand or even a few million on the market. Land grabs and demolition had already become some of the most heated issues among the public at the time.

In most cases of anti-incinerator protests, there were pre-existing conflicts of interest about land. This was true in the case of the Xidong waste incinerator conflict in Wuxi, and the Likeng waste incinerator conflict in Guangzhou. There were basically two kinds of situation. In the first, the people who commandeered the land 'owed' the original villagers in some way: either the compensation price was on the low side (this was especially true with early projects), or part of the money did not reach the villagers promptly, or corrupt cadres took it. The other situation was when people whose land was being commandeered were trying to get a better deal. They

24 After the 2009 International Symposium on Dioxin, frictions and conflicts against waste incinerators increased rapidly, which generally proves the above point about public grievances. For a summary of conflicting events against waste incinerators and waste-energy power plants, see Chen, Wu, et al., *Dilemma and Way-Out of Urban Household Waste Disposal*, pp 86-90.

25 Unlike in the West, CCTV programmes can often affect public attitudes and in some cases even provide legitimacy for rash behaviour. The public thinks that if something is broadcast by CCTV it must to some extent reflect the opinion and position of the government. In the years before and after 2009, CCTV had a relatively open and progressive attitude towards environmental issues, while local government and enterprises were relatively backward. The coverage of the channel is very wide, and so ordinary people often used the views expressed in its programmes as the basis for opposing environmental pollution.

26 The following analysis of the anti-incinerator movement comes mainly from Chen, 'Walking Out of the Dilemma of Waste-Energy Incinerator Plants'.

hoped that by bringing complaints they could get a more favourable policy or put themselves in a better position to negotiate over the terms of the land acquisition. This kind of tactic was especially obvious in places where the market mentality was stronger. A further issue was that when land was appropriated for other kinds of projects, these ventures nearly always brought opportunities for local development, such as new employment opportunities. But with incinerators, there were almost no positive benefits for the village. Instead they brought a lot of negative impacts, including pollution, health risks, falling real estate values, ruined views, and psychological effects.

In anti-incinerator incidents, 'baseline' pollution was nearly always an underlying problem. 'Baseline' pollution refers to the pollution situation in the area prior to the building of the incinerator and to local residents' perception of such pollution. These were major contributing factors in the attitude of local people towards incinerator projects. From an objective perspective, if pollution in the area was already very serious, adding a new waste incinerator would overwhelm the area's 'environmental carrying capacity'. And in terms of the residents' ability to tolerate pollution, adding more to what they had already been experiencing would make them feel they could not stand it anymore, and lead them to take action in opposition. This is what happened in Likeng in Guangzhou, where the residents still vividly remembered the noxious impact of the earlier landfill. That dump had a relatively limited life span, and the residents hoped that the environment would gradually recover, but once the waste incinerator was built, they felt there would be no end to it, and the risk to their health would be there forever. This was a major challenge to the villagers' tolerance. If landfills already presented a conflict of interest for residents and a problem of background pollution, then waste incinerators added the extra element of dioxin, and this served as a tipping point. This is why incinerator projects generated an even stronger wave of public opposition than landfills.

Dioxin was a perfect topic to become the object of rumours, for which Allport and Postman (1947) summarized the conditions as 'importance and ambiguity', the latter including information that is either lacking or unclear.[27] Dioxin obviously scores high on both these measures. Its high toxicity and strong carcinogenic nature undoubtedly make it extremely important but (at least at this time) its impacts were also unclear in many ways. Due to limitations of technology and the high monitoring cost, waste incinerators did not have online emission-monitoring systems and there was no other standardized system for monitoring dioxin emissions. As a

27 Allport and Postman, *The Psychology of Rumor*, pp. 501-507.

result, even experts and local government officials did not know what the exact situation was. There were also differences of opinion among experts about the safety of waste incinerators. The majority of technical experts considered that dioxin could be controlled, but some experts disagreed, and at the time an effective system for regulating dioxin was not in place.

When the media broadcast information about the health risks associated with dioxin, some scientific knowledge was imparted, including the fact that incinerators produce dioxin, and dioxin is highly toxic. But the media also spread 'edited stories' that were tailored to the local audience. What is more, some people deliberately spread rumours, like the fake 'List of Cancer Patients in Yongxing' in the Likeng incinerator case. In this incident, the number of people with lung and other respiratory tract cancers was almost doubled in order to construct a causal relationship between pollution from the incinerator and high rates of respiratory tract cancer in the local population and attempt to provide a 'scientific basis' for opposition to the incinerator (see Chapter 8).

Revolutionary changes in information and media technologies were very relevant to the explosion of the 'cancer village' phenomenon. The development of online media led to a huge increase in the volume of information in circulation. At the same time, people could now access the internet on their mobile phones, and functions that enabled people to post and send messages to whole groups meant that it was possible to spread information anonymously to a large number of people in a very short period of time. In this explosion of information, apart from deliberate rumour mongering, it was also common for stories to be distorted in the retelling, or misinterpreted.

The Post-Cancer Village Era

After a decade of being a hot topic, the issue of 'cancer villages' slowly lost its heat. However, understanding regional variation in the incidence of cancer as an environment-health problem will require a much longer process of scientific enquiry.

In the last five years (2012-2017), there has been a big shift in the situation regarding pollution control in China: the country is in the process of upgrading its economic structure, and environmental management has improved considerably. In November 2012, in a report at the eighteenth Party Congress, the government introduced the idea that 'economic construction, political construction, cultural construction, social construction, and the construction of ecological civilization are five parts of one united objective'. Historically, from 1949 to 1978, China emphasized 'politics in the lead'; and after 1978 it emphasized 'putting economic construction at the centre'. In contrast, the idea of 'Five Parts in Unison' (*wuwei yiti*), emphasizes integrated progress across the whole system. Local governments have also been actively promoting pollution control, as in the 2013 plan in Zhejiang which highlights 'dealing with five kinds of water problems at the same time' (cleaning up polluted water, preventing floods, draining waterlogged fields, ensuring the water supply, and saving water). Meanwhile, Beijing is making an ever-fiercer effort to deal with its air pollution and claims to have invested 760 billion RMB in controlling PM2.5 (Particulate Matter 2.5).[28]

In terms of environment and health, some specific measures are also being implemented. Whether it is industrial pollution or agricultural pollution or household waste, water is an important medium through which pollution affects people's health. Overall, with rising living standards, tap water coverage has risen. In 2003, the Ministry of Health published the 'Main Results of the Third Survey of National Health Services', which showed that tap water availability had reached 96% in urban areas, 'but in rural areas the coverage rate is only 34%, and in some poor areas, the proportion of households without safe drinking water is still quite high'.[29] After the twelfth Five Year Plan (2011-2015), it reported that 'the proportion of rural households benefiting from collectively provided water rose from 58% in 2010 to 82% in 2015, and the tap water coverage rate in rural areas reached 76%. There was also a clear improvement in the quality of water provided'.[30] When some places encountered pollution problems, they generally turned to improving the water supply as a way to resolve them. For example, after the high incidence of cancer was reported in Huangmengying Village, the government took measures to provide the villagers with deep wells, because although the surface and shallow ground water was polluted, the deep ground water was not, and could serve as a source of clean drinking water. Clearly, improving the water supply is an important way to lower the risk of cancer.

After 2009, widespread protests against waste incinerators brought about a temporary slowdown in the construction and operation of incinerators.

28 See News China.com, 'Beijing Municipal Government Made a Vital Promise to the Central Government to Control Air Pollution, 760 Billion RMB Will Be Invested to Control PM2.5'.

29 Zhu and Zhou, 'China's Tap Water Coverage Rate is 96% in Urban Areas and 34% in Rural Areas'.

30 Yu, 'China's Tap Water Coverage Rate in Rural Areas Will Reach Above 80% by 2020'.

And the reflection that took place during this period has had a positive impact on the later construction and operation of waste incinerators. In later construction projects, more caution was exercised in choosing sites for incinerators. The environmental impact of projects was taken seriously, and awareness of health risks rose. Most new waste incinerators were sited in places that were far from population concentrations, and more mature, safer, and more reliable technologies were chosen. There was also a stronger emphasis on controlling pollution and health risks. After going through many episodes of conflict, the relationship between the public and local government was adjusted, and the public mood gradually became calmer.

The heat gradually went out of the issue of 'cancer villages', but cancer is constantly undergoing change. As average life expectancy gets longer, the likelihood of developing cancer increases. Furthermore, with changes in lifestyle, the incidence of chronic diseases, including cancer, rises. People's awareness of cancer prevention has increased and the possibility that a single pollutant will cause a concentrated outbreak of cancer has fallen; but accumulated pollution, and the risk of it combining with other factors to cause cancer is still on the rise.

Works Cited

Allport, Gordon W., and Leo Postman. *The Psychology of Rumor.* New York: Henry Holt, 1947.

CCTV (China Central Television) Programme, 'International Symposium on Dioxin Finds Waste Disposal Encountering Problems, and Dioxin is Troubling China.' 1 September 2009.

Chen, Ajiang. 'Walking Out of the Dilemma of Waste-Energy Incinerator Plants', *China Social Science Today*, 21 March 2014.

—. *Secondary Anxiety: A Sociological Perspective of Water Pollution in the Tai Lake.* Beijing: China Social Science Press, 2009.

—, Wu Jinfang, et al., *Dilemma and Way-Out of Urban Household Waste Disposal.* Beijing: China Social Science Press, 2016.

China Industry Information Net. 'General Analysis of China's Fertilizer Industry's Financial Operation', 3 July 2014. http://www.chyxx.com/industry/201407/261061.html. Accessed 5 February 2020.

Guan, Yanyan, Xu Xiaoming, Jin Guangqing, Lai Zhihui. 'Using ICP-AES to Test the Level of Chemical Element Cu, Zn, Fe and Mg Contained in the Whole Blood of People Who Are Under High Exposure to Minerals', *Contemporary Preventive Medicine* 36, no. 19 (2009): 3706-3707.

Ministry of Environmental Protection, State Statistical Bureau, and Ministry of Agriculture. *Report on China's First Pollution Census.* 2010. http://www.stats.gov.cn/tjsj/tjgb/qttjgb/qgqttjgb/201002/t20100211_30641.html Accessed February 25, 2010.

National Bureau of Statistics. *China Statistical Yearbook 1985*. Beijing: China Statistics Press, 1985.

National Office of Tumour Treatment and Prevention Research. *National Survey of Mortality from Malignant Tumours 1990-1992*. Beijing: People's Health Press, 2008.

News China.com, 'Beijing Municipal Government made a Vital Promise to the Central Government to Control Air Pollution, 760 Billion RMB Will Be Invested to Control PM2.5,' 19 Jan 2014. http://news.china.com.cn/2014-01/19/content_31236266.htm. Accessed 5 February 2020.

Pei, Linzhi. 'Experts Talking About How to Improve the Utilization Rate of Fertilizers.' *Agricultural Technology*, 1999(1). 10-11, 14.

People.cn. 'Our Country's Mortality Rate from Lung Cancer Increased Almost Ten Times Within a 40-Year Period,' 18 March 2013. http://scitech.people.com.cn/n/2013/0318/c1007-20819577.html. Accessed 5 February 2020.

Sun, Shiquan, Meng Xianyu, Yuan Liyun, You Zhanyun, Liu Shengen, Yang Xiaoou, Yang Lan, Chen Hongyu. 'Etiological Analysis of Yunnan Gejiu Tin Miners' Lung Cancer.' *Radiation Protection,* 1981 (2). 1-8, 12.

WHO website, 4 October, 2016: https://www.who.int/en/news-room/fact-sheets/detail/dioxins-and-their-effects-on-human-health. Accessed 5 February 2020.

Xu, Zhixiang, ed. *Fertilizer, Waste Water and Oesophagus Cancer.* Beijing: Science Publishing House, 2003.

—, Tan Jiaju, Han Jianying, and Chen Fenglan. 'The Nitrogen Cycle Hypothesis and the Cause of Oesophagus Cancer, Stomach Cancer and Liver Cancer.' *Medical Research Magazine* 1 (2008). 61-74.

Yang, Gonghuan and Zhuang Dafang, eds. *Atlas of the Huai River Basin Water Environment: Digestive Cancer Mortality.* Dordrecht: Springer. 2014.

Yu, Wenjing. 'China's Tap Water Coverage Rate in Rural Areas Will Reach Above 80% by 2020.' *State Council, People's Republic of China*, 2016. http://www.gov.cn/xinwen/2016-01/12/content_5032318.htm. Accessed 5 February 2020.

Zhu, Haifeng. 'Waste Mining Water Harmed the Whole Village.' *China Mining Newspaper,* 20 January 2001.

Zhu, Yu, and Zhou Tingyu. 'China's Tap Water Coverage Rate is 96% in Urban Areas and 34% in Rural Areas', *H2O China*, 3 December 2004. http://www.h2o-china.com/news/32787.html. Accessed 5 February 2020.

2 The Ins and Outs of a 'Cancer Village'*

Chen Ajiang

Abstract
This chapter discusses a research site in the Huai River Basin, using the perspective of 'ins and outs' as an analytical framework. The pollution in Huangmengying Village mostly comes from upstream, but water conservancy projects, the nature of water courses and flows of pollution also have an impact. Media coverage of cancer in the area emphasized the role of external pollution from industry, but this study also found a relationship between smoking and cancer among men and between hepatitis B and liver cancer. There was also an association between drinking water and the high incidence of digestive tract diseases, which declined when the water supply was improved. The phenomenon of 'cancer villages' is therefore also shaped by the living conditions and lifestyle of the villagers.

Keywords: water courses, environment and health, lifestyle habits, internal pollution, external pollution

Introduction

If the word 'cancer' refers to a condition that affects only personal physical health, the term 'cancer village' implies a problem that goes beyond the individual. This is because the term 'village' refers not only to the physical and social location in which a community of people resides, but also to the community itself. In pre-modern times, rural areas of China were mainly comprised of villages and so this social unit has special significance. The phenomenon of 'cancer villages' has therefore become a subject of debate not

* Some of the material in this chapter was previously published in Issue 2, 2013, *The Journal of Guangxi University for Nationalities* (Philosophy and Social Science Edition), and Issue 6, 2012, *The Journal of Chinese Social Science* (Inner Manuscript Edition). Some names have been changed.

Chen Ajiang, Cheng Pengli, and Luo Yajuan, *Chinese "Cancer Villages": Rural Development, Environmental Change and Public Health*. Taylor & Francis Group 2020
DOI: 10.5117/9789089647221_CH02

only among residents of villages themselves, but also among outsiders and even the national government. 'Cancer villages' have also become a topic of social science research. At the same time, the problem of cancer villages can be understood from many different perspectives. These can be simply grouped into internal and external factors, in the sense that they relate to whether pollution is generated from within or outside the community and to whether illness is caused by individual behaviour or the impact of external factors, including pollution.

In medical science, it is generally considered that the risk factors for cancer include: (1) Genetic factors. The probability of getting certain types of cancer varies widely across racial groups, and men and women are prone to different cancers. (2) Diet and nutritional status. Long term deficiency in certain nutrients, such as selenium, is positively correlated with several types of cancer, including colon, colorectal and prostate cancer. Recent studies also show that being overweight or obese may cause pancreatic cancer and breast cancer. (3) Viruses and bacteria. Although there is no conclusive evidence showing that viruses directly cause cancer, epidemiological surveys find that 15-20% of cancers can be related to viruses and biological factors. For example, Hepatitis B has been found to be a precursor of primary liver cancer. (4) Lifestyle factors. Smoking causes lung cancer, and it has been found that eating too fast and eating very hot food may increase the risk of oesophageal cancer. (5) Environmental pollution.[1] Genetic and lifestyle factors, including viruses, nutrition, and individual behaviour, can be regarded as internal factors, while environmental pollution can be seen as external. This study examines the phenomenon of cancer from these 'internal' and 'external' perspectives, taking the village as the unit of analysis.

In 2004, China Central Television (CCTV) broadcast an influential programme, titled 'Rivers and Villages', which told the story of Huangmengying Village in Henan Province. It described Huangmengying as a 'cancer village', and focused primarily on water pollution from outside the village. The programme highlighted the previously unrecognized issue of 'cancer villages', and this attracted the attention of the central government and led to studies investigating the high incidence of cancer and its relationship to pollution. Yet, my own later fieldwork in Huangmengying found that other 'internal' factors were also likely to be contributing to the high rate of cancer there, including smoking among men (lung cancer); high rates of Hepatitis B (liver cancer) and drinking 'raw' (unboiled) water (associated with high rates of intestinal diseases). This chapter explores the relationship between

1 Wang, *Liu Xing Bing Xue* (*Epidemiology*). pp. 284.

village residents' lifestyles and their diseases. It reveals the complex causes of cancer and questions the single focus of the public and the media on polluted water as the sole cause of cancer in Huangmengying.

This chapter draws on primary data from two rounds of fieldwork. The first study was conducted in April 2009, when Huangmengying was studied as a negative example in the project, 'Harmony between People and Water'. The second field trip was conducted in August 2011. This follow-up investigation was prompted by the results of a study of several cancer villages in South China sponsored by the Social Science Research Council in 2010. Secondary data are drawn from the videos and written materials of the CCTV programme 'Rivers and Villages', as well as literature on pollution and disease in the Huai Basin from the China Knowledge Resource Integrated Database (CNKI).

Mr Wang, the village doctor, was a key informant for the research in Huangmengying. He was very knowledgeable about pollution and disease in the village. Dr Wang had basic medical training and considerable clinical experience: he had been practicing for over twenty years and often saw more than 30 patients per day. In addition, he was born and raised in the village, which was a typical 'familiar society' (*shuxi de shehui*) in which everyone knows everything about everyone else,[2] including their economic status, marriage and family relations, local customs and lifestyles. It is therefore not surprising that this village doctor was very knowledgeable about disease, and especially about the diagnosis, treatment and prevention of chronic diseases that were closely related to the economic, social and cultural background of his patients. Lastly, Dr Wang was a member of Huangmengying's elite. He had formerly been the village accountant, and this, combined with his special position as village doctor, meant that he understood the public affairs of the community. With the assistance of Dr Wang, in August 2011 we updated the list of patients who died of cancer shown in CCTV's programme in 2004. In the second investigation, the research team also interviewed experts from the County Centre for Disease Control (CDC), the Environmental Protection Bureau (EPB), and the Provincial Department of Water Resources, Division of Agricultural Water. The Provincial CDC and the Institute of Chronic Non-communicable Diseases were also interviewed for background and for information about pollution and disease.

2 Fei, *From the Soil.* pp. 57-59. The term that is often quoted is 'shuren shehui' (translated as the 'acquaintance society') but in fact Fei Xiaotong says 'Because people seldom move, they grow up not only among familiar people (*shuren*) but also in familiar places (*shuxi de difang*)'. We use the term 'familiar society' (*shuxi de shehui*) to cover both people and place. This apparently minor adjustment is actually very significant for environmental sociology.

Literature Review

When we began this research, 'cancer villages' had become a hot topic in the media. A search on the Chinese search engine Baidu on 30 April 2012 using the term 'cancer village' produced 835,000 results. A similar search on Google had about 2,650,000 hits. It is also notable that search engines and internet encyclopaedias such as Baidu Baike, Wikipedia, and Interaction (*hudong baike*), all have special entries for 'cancer village'.

Yet although the term 'cancer village' was widely used, it was never clearly defined. While it may appear to refer to villages that have a relatively high incidence of cancer morbidity or mortality, the reality is more complex. The designation of villages in China as 'cancer villages' resulted not from the findings of medical research, but mostly from accusations by outsiders (especially the media) or villagers' own claims. From a natural science perspective, the term 'cancer village' is therefore often problematic because it is not supported by adequate data or rigorous investigation. Looked at from a social science perspective, we see that the social construction of cancer villages was related to the particular situation at the time, when environmental pollution was worsening, concern about health risks was increasing, and there was a lack of social trust. It is very hard to determine who first came up with the term 'cancer village', and perhaps it is not really important who was the first to coin it; the fact that the term came to be publicly accepted and widely used is already a sufficiently complicated phenomenon.

As noted above, the village has been the most important spatial unit of life in China. Because of this, it has been common to put an adjective before the word 'village', to modify or describe it. For example, some villages use a clan or family name as a modifier, such as Wang Family Village (*Wang Jia Cun*) or Li Family Village (*Li Jia Cun*), even though these surnames are common all over the country and not specific to that village. Other villages have names that refer to local scenery such as mountains, rivers, grasses or trees, or to ethnic groups or historical events. Some village names also reflect recent changes in Chinese society. For example, newly developed areas of certain cities have been named 'New Mine Village' or 'Electronic Village', and urban areas settled by migrants from a certain region – known as Villages in the City (*chengzhongcun*) have been named 'Zhejiang Village' or 'Henan Village'. There is also a history of naming villages formed by clusters of people isolated due to medical or psychosocial problems, for example, areas where leprosy or HIV sufferers live together might be called 'Leprosy Village' or 'HIV Village'. It seems that 'cancer villages' belong to this last type, where the village's identity is constructed around the presence of a shared disease.

A review of literature in the CNKI database found that the term 'cancer village' first appeared in the *China Mineral News* on 20 January 2001. An article by Zhu Haifeng, titled 'A Village Polluted by Mining Wastewater',[3] reported on a village in Guangdong Province that had a high incidence of cancer. Zhu claimed that historically this was a beautiful village with plentiful fish and rice, but that it had since become notorious for poverty and cancer. In the same year, the journal *Environmental Protection* published an article by Wang Shaofang and Lin Jingxing, titled 'The Ecological Environment and Geological Disease: An Empirical Analysis of a "Cancer Village" in Shanxi'. Through an analysis of water, soil, rock, flower and vegetable samples taken from Longling Village in Hua County of Shanxi, as well as analysis of human hair samples, the authors argued that arsenic, lead and cadmium were the major causes of cancer in the village. These two articles were the earliest to appear in the CNKI Database. Both were in professional journals and they attempted to analyse the cause of cancer from a scientific perspective.

A literature review of two other relatively systematic and authoritative sources within the CNKI – the China Academic Journal Web Publication Database (Journal Database) and the China Important Newspaper Full Text Database (Newspaper Database) – found that from 2001 to 2011, 52 journal articles and 60 newspaper articles on 'cancer villages' were published.[4]

The data in the table below show that starting in 2004, there were at least ten articles about 'cancer villages' every year.

Content analysis of this literature found that no article gave an official definition of the term 'cancer village'. However, despite this, the use of the term implied some important shared assumptions. These included the assumption that the incidence of cancer or the mortality rate among village residents was unusually high; that unusual cancer cases appeared in one or several villages; and that these unusual cancer cases were due to pollution in or around the village (even if these claims were not confirmed by a government agency or formal scientific research). In most cases, the

3 The term 'cancer village' probably appears earlier in other internet literature databases, but as these are constantly changing and being updated it is difficult to compare them over time. CNKI is not only authoritative, but also stable.

4 As of June of 2011, the China Academic Journal Network Publishing Database included more than 7,700 journals, and a total of more than two million texts.

http://acad.cnki.net/Kns55/brief/result.aspx?dbPrefix=CJFQ

Table 2.1 Number of articles about 'cancer villages' in journals and newspapers, 2001-2010

Year	*Journal*	*Newspaper*	*Total*
2001	4	1	5
2002	0	0	0
2003	0	2	2
2004	7	12	19
2005	6	8	14
2006	5	4	9
2007	9	7	16
2008	8	3	11
2009	5	12	17
2010	8	11	19
Total	52	60	112

Data Source: Separate searches of the Journal and Newspaper literature database in CNKI using the term 'Cancer Village'

assumed sources of pollution were industrial or mining facilities. Some articles not only made these implied assumptions but also proceeded to make recommendations about how the government could deal with the 'cancer village' problem.

Third, if the articles are sorted into the categories of natural science, social science (mainly law and sociology) and media reports, it is quickly apparent that most articles on 'cancer villages' were news reports or media discussions about reports of 'cancer villages'. The first report on a 'cancer village' appeared in 2001. Then there was a silence for a while between 2002 and 2003, followed, from 2004 to 2005, by a rapid increase in the number of articles. In addition to follow up reports on the villages in Guangdong and Shanxi village that had already been reported on earlier, a huge literature on 'cancer villages' emerged in 2004-2005. This had some new characteristics. There were very few medical science articles and the number of environmental science papers also decreased, while the number of newspaper articles, and economics and rural studies articles increased. Most of the authors worked in the media and most of the studies were simple investigations and observations, not academic research conducted by scholars with a background in environment science, chemistry or medical science. In terms of their spatial distribution, most of the cases were in the Yangtze River and Huai River Basin regions, which were both seriously polluted parts of the country.

Water system change and the characteristics of regional pollution

Huangmengying Village is located three kilometres south of the Shaying River in eastern Henan Province, and 30 kilometres west of the border between Henan and Anhui. The Shaying is the largest tributary of the Huai River. It originates in the Funiu Mountains in Henan Province and flows through more than 40 settlements in Henan and Anhui provinces, including Pingdingshan, Luohe, Xuchang, Zhoukou, and Fuyang. The Shaying is over 600 kilometres long and has a basin area of nearly 40,000 square kilometres. In the 1990s, water pollution in the Shaying was much more severe than in other rivers in the Huai Basin.

To understand the situation in Huangmengying, it is helpful to understand the changes that have taken place over the course of time in the river on which the village is situated. Historical changes in the river system itself, along with later water engineering projects, transformed the entire water system, and these changes, in turn, affected the ways in which the water pollution that occurred with industrialization affected the region. Furthermore, once the Yellow River changed its course and began to flow into the sea via the Huai River, this caused upheaval in the water system and natural disasters, which hampered the economic development and social integration of the Huai River Basin. These natural and socioeconomic factors are the backdrop to the emergence of 'cancer villages'.

There is a popular saying that goes, 'No matter how many places you travel to, the banks of the Huai are the best'. This reflects the fact that the Huai Basin was once a prosperous region; during the Song Dynasty and earlier it was noted by visitors for its wealth and beautiful scenery. Yet in the 2000s, the area was famous for disasters, poverty, and pollution. How did this transformation occur? As mentioned above, it was partly due to changes in the river system that occurred without human intervention. Before 1128 AD, the Huai had its own estuary and it was a wide, deep river that flowed without obstruction into the sea. Everyone is familiar with the Dujiangyan irrigation system, constructed in 256 BC, which enabled the Chengdu Plain in today's Sichuan Province to become a land of abundance, free from floods and drought. A similar ancient water engineering project, Quebei, was conducted 2,600 years ago in the Huai Basin and enabled the irrigation of a large area of farmland. Along with other, similar projects, this helped agriculture to flourish. Because of its advantageous location between the Yellow River and the Yangtze, the Huai Basin also had a well-developed transport system. The Han Canal, built in 486 BC, connected the Huai and the Yangtze, and the Hong Canal, built in 360 BC, linked the Yellow River and several tributaries (the Danshui, Suishui, Shashui, and Yingshui) of the Huai, forming a water transportation network linking the two major rivers.[5]

However, in 1128 AD, the Yellow River changed its course, turning southwards, and for the following 700 years it took over the Huai as its route to the sea, switching its main course back and forth between various tributaries (the Yingshui, Guoshui, Suishui and Sishui). To make matters worse, the Yellow River carried a lot of silt, which built up in the lower reaches of the Huai where it fed into the sea, impeding the flow of water in the low-lying middle and northern reaches and causing it to back up and flood the region.[6] The problem was exacerbated by the climate. The Huai Basin is located in a transitional area between north and south China, which means that it is dry in winter and spring, and rainy in summer and fall, with both droughts and floods common. People in the region used to say that 'Heavy rain causes heavy floods; light rain causes light floods, and no rain causes drought!' For this reason, until the government took action after 1949, the Huai Basin and the areas of Henan and Anhui that it flowed through were prone to all kinds of natural disasters.

Starting in the 1950s, the national government launched a comprehensive set of measures to address water-related problems in the Huai Basin, tackling both flooding and the need to irrigate farmland with water engineering projects in the middle reaches of the river that enabled water to be stored at certain times and drained at others. After many years of hard work, the Huai Basin became one of China's major grain producing regions. Compared with the past, it was free from disasters and floods. People's basic needs for food were met and the region developed quickly. Nonetheless, it continued to lag behind other areas of China, and despite the enormous changes that have taken place, Shenqiu County where Huangmengying is located is still a state-level 'poverty county' that is over-populated and has a very poor industrial and commercial base.[7] At the time of our research, farmers earned most of their income by migrating to cities for work.

The water engineering projects conducted in the Huai Basin in the 1950s and subsequent years transformed the original river system, and this affected the later distribution and concentration of pollution. To take

5 Zheng, *The Water Conservancy Encyclopedia of China: The History of Water Conservancy*, p. 28.

6 Ibid., p. 29.

7 In 2012 the Development Office of the State Council Leading Group on Poverty Alleviation published a *List of Key Counties for National Poverty Work*. Shenqiu County in Henan was among the 592 counties on the list. http://www.cpad.gov.cn/art/2012/3/19/art_50_23706.html

Huangmengying and the surrounding area as an example, the Huaidian Sluice Gate retains water from the upper reaches of Huai, raising water levels and effectively forming a reservoir. The Shaying River, which is above the gate, flows through the main and sub-channels of the Shanan River into Huangmengying and the surrounding farmland. Run-off from the farmland and villages then flows back into the lower reaches of the Shaying. The result of this is that if a lot of pollution is released into the upper reaches of the system, it spreads to villages and farmland downstream. As it flows through fields and villages, the river also picks up nutrients from fertilizer and human and animal waste, and these concentrate in the river and flow into other settlements and fields further down. The situation of the Shaying River is shown in Photograph 1 and Photograph 2.

Field research and the literature show that water pollution in Huangmengying came from three sources. The first and primary source is industry in the upper reaches of the Shaying River, including MSG (Monosodium Glutamate), paper, and leather factories. For example, in one of our earlier studies, we investigated the Maoji factory, which is typical of the leather industry, a highly polluting sector.[8] However, the impact of pollution from these factories was also related to the characteristics of the river system and water flows discussed above. Several incidents demonstrate these relationships.

In 1994, an incident occurred in which pollution affected a 70 kilometre stretch of the Huai River from the Shaying in Henan to Hongze Lake in Jiangsu. Some places in the affected area had to stop using drinking water, some fisheries lost a lot of their stock, and some factories were shut down. The cause of the problem was that due to a persistent drought, most sluice gates in the Huai Basin had been closed, causing pollution to concentrate. Then, in July, there was heavy rain. On 14 July, the Shaying Sluice Gate was opened to release the floodwater and the polluted water flowed into the mainstream of the Huai River, and then on down into the Huainan and Bengbu municipalities. On 19 July, the Bengbu Sluice Gate was opened, and two hundred million cubic meters of polluted water flowed into the lower reaches of the Huai River, forming a huge pollution plume 70 kilometres long.[9]

Subsequent heavy pollution incidents in the Huai River in 1995, 1999, and 2004 were very similar. In fact, after economic reform and the development

8 Chen, Luo and Chen, 'The Paths and Consequences of "Pollution First, Clean Up Later"– A Case Study of Environmental Protection and the Maoji Leather Industry'.

9 Song, ed. *China Huai River Basin Water Environment Protection Policy Evaluation*, p. 416.

of industry along the Shaying, the quality of water in the river worsened steadily. The Shaying was the major water source for agriculture in Huangmengying and as the river was polluted, the village and the area around it were inevitably affected too. Polluted water from the Shaying flowed into farmland around the village via its main and subsidiary channels, which were also connected with ponds in the village. Moreover, many villagers' shallow wells were close to these ponds and the polluted water seeped underground and polluted them. In 2004, CCTV tested three samples of water from the river, and the main channels and ponds in Huangmengying. Five indicators, including Chemical Oxygen Demand (COD) and ammonia nitrogen, were above the standard for Grade V water, which means that they were unfit even for agricultural purposes.[10]

The second source of pollution came from within Huangmengying itself. When analysing the causes of water pollution in the Tai Basin, I found that when the water in an area was polluted by external sources for an extended period, local residents sometimes stopped taking care of the water and engaged in polluting activities themselves.[11] Later, when I conducted research in other places in China, including Northern Jiangsu, Northern Anhui, and Eastern Henan, I again found that villages affected by industrial pollution also suffered from severe pollution from within the village. There was a similar situation in Huangmengying. When industrial pollution became severe, residents started to throw their household garbage into the ponds (see Photograph 3). When we visited the village in 2009 and 2011, there was still a lot of garbage around the ponds, even though since 2004 the village had had several years of environmental education and propaganda and had implemented various environmental governance and improvement programmes.

Third, water quality problems may have been partly due to the presence of unusual chemicals resulting from geological factors. According to the tests conducted by CCTV, water in one well owned by the Sun family, which was eight meters deep, had excessive levels of manganese. Viewers of the programme were given the impression that the water in the well exceeded the standard due to pollution from the Huai River, but according to the literature and my own fieldwork, it may have been due to unusually high naturally occurring levels of the chemical. The document 'National Engineering for the Safety of Rural Drinking Water in the 11th Five-year Plan', contains a chapter called 'the Current Situation and Main Problems

10 CCTV, *River and Village*.

11 Chen, *Secondary Anxiety: The Social Interpretation of Water Pollution in Tai Lake Basin*, p. 132.

with Rural Drinking Water Safety' which includes a section titled 'Other Polluted Drinking Water: Distribution, Causes and Damage'. It states,

> Across all rural areas of the country, the problem of iron and manganese in excess of national standards affects 44.1 million people, accounting for 14% of the population who have unsafe drinking water and 19% of those affected by unsafe water quality.
> The problem of excessive iron and manganese is caused mainly by the geological structure and hydro-geological conditions [...] Excessive iron, manganese and heavy metals are also caused by human behaviour, including random dumping of waste from mining and smelting and unprocessed waste water flowing into surface water or polluting groundwater.[12]

Judging from the geological conditions and the evolution of the river system in Huangmengying, the excessive manganese might have been caused by the underlying geological structure and hydro-geological conditions. As far as I could determine, factories in the upper reaches of the Shaying did not have large scale industry or mining that involved manganese. Furthermore, if the problem had been caused by pollution from the Shaying, then one would have expected water in other wells in the village to contain excessive levels of this element, but they did not. We found a similar situation in Jiangxi, where manganese levels in the wells of some residents exceeded standards while in others it was normal.[13] Considering the situation as a whole, it seems that excessive manganese was more likely to be caused by the underlying geology of the region than by pollution.

Pollution from outside the village and cancer

By the mid-1990s, the problem of pollution in the Huai Basin was very serious and attracted the attention of the central government. In May 1994, the State Council conducted an inspection of the enforcement of environmental law in the region, and in August 1995, it issued the 'Interim Regulations for Preventing and Controlling Water Pollution in the Huai Basin', the first

12 National Development Reform Committee, *National Rural Drinking Water Safety Project – The 11th Five-Year Plan.*

13 Chen and Cheng, 'Risk Perception and Responses to the "Cancer-Pollution" Problem – Empirical Studies in Several "Cancer Villages"'.

regulation for the prevention and control of water pollution in a river basin in China. In 1996, the State Council proceeded to approve the 'Plan for the Prevention and Control of Water Pollution in the Huai Basin during the Ninth Five-year Plan'. During the Ninth Five-year Plan, the government focused on improving the water environment of 'three rivers and three lakes'. The Huai River was one of them.

Although the government and the public had high hopes for the improvement of the water environment in the Huai, problems with pollution recurred in 2004. In the middle of July, heavy rain in the upper and middle reaches of the river forced local governments to open sluice gates in these sections. A lot of polluted water entered the mainstream of the Huai River, and a body of waste water over 150 kilometres long drifted downstream, calling the effectiveness of pollution control in the river into question.[14] It was in this context that the TV programme 'Rivers and Villages', broadcast in August 2004, raised the acutely sensitive issue of whether water pollution in the Huai Basin had already damaged public health, and people were shocked by images suggesting that water pollution had caused a high incidence of cancer.

CCTV conducted an investigation in Huangmengying and gathered a lot of basic information. They also asked a qualified research institute to test water samples and consulted some experts. The following four paragraphs describe the way in which the issue of disease and the relationship between disease and pollution was discussed in 'Rivers and Villages'.[15]

The high incidence of cancer was immediately introduced as a severe problem. When the journalist walked into the village, there was a funeral going on. She said, 'Huangmengying Village has been famous for its cancer patients for more than a decade. This year, seventeen names were added to the list and eight of them have already died'. Accompanied by the Secretary of the Village Communist Party Branch, the journalist walked along a street in the village and said, 'In this street, eight families in a row have cancer patients'. She continued, 'According to the Village Committee's records 204 people died in a fourteen-year period between 1990 and 2004, and 105 of them died of cancer, or 51.5% ... Obviously, cancer is prevalent here. Most people who died of cancer were about fifty years old, but the youngest was only one year old'. At the end of the programme, a list of dead cancer patients' names was displayed.

14 Lin, 'Game Theory Analysis, Relationships and Patterns: An Example of the Prevention of Pollution in the Huai River Basin', pp. 342-343.
15 CCTV, *River and Village.*

What caused the high incidence of cancer in this peaceful village? The CCTV journalist analysed several possible causes and excluded first food and then air pollution as possibilities. She eventually concluded that water pollution was responsible, saying, 'If the high incidence of cancer is not caused by food and air pollution, the most likely cause is polluted drinking water'. According to the Party Secretary, Mr Wang, the nearer villagers lived to ponds and irrigation ditches, the higher the incidence of cancer was, while it was lower among those living further away from these water bodies.

The relationship between water pollution in the Huai Basin and pollution in wells was portrayed as obvious. When water in the Shaying River was polluted, it flowed into farmland and villages via irrigation ditches. For example, the Huaidian Sluice Gate on the Shaying River held 200 million cubic meters of water to irrigate 750,000 *Mu* of farmland, supplying more than 500 villages.[16] Polluted water flowed into irrigation ditches and ponds around these villages, where the wells were close to ponds and only a few to a dozen meters deep. These wells quickly became polluted. According to tests of water samples from a pond in Huangmengying, five indicators, including COD and ammonia nitrogen, were above the Grade V standard. The CCTV programme stated that 'Since polluted water from the Shaying River has flowed into Huangmengying, the water in the ditches and ponds has become darker and darker, fish and prawns have more or less died out, and the number of cancer patients and deaths has risen'. This was the basic conclusion of the programme.

From the perspective of medical science, it is very difficult to determine the cause of the high incidence of cancer. Yet, once the use of the term 'cancer village' became widespread, it had a massive ripple effect, and a profound social impact. The TV programme soon attracted the attention of the central government. After the Huangmengying 'cancer village' was reported in the media, Prime Minister Wen Jiabao immediately ordered the Secretary of the Henan Provincial Party Committee, Li Keqiang, and the Henan Provincial Governor, Li Chengyu, to investigate the problem of the high incidence of cancer in the Huai Basin. Later, the State Council organized the four provincial governments of Henan, Anhui, Jiangsu, and Shandong and the relevant ministries to launch a project for the prevention and control of pollution and cancer in the Huai Basin[17]. In 2004, the Ministry of Health issued the 'Programme for Cancer Prevention and Control in

16 1 *mu* = 667 m^2.

17 Lu, 'The Orientation of Cancer Prevention and Control in China as Viewed from the Development of Esophageal Cancer Scene'.

China (2004-2010)', which laid out the guiding ideas, objectives, tasks, and strategies for cancer prevention and control. In 2005, the Ministry of Health initiated Central Transfer Payments for the Early Detection and Treatment of Cancer.

In 2005, the Chinese Centre for Disease Control and Prevention invited the Chinese Academy of Medical Sciences, Peking Union Medical University, and other universities and research institutes to participate in a project titled, 'an Investigation into Malignant Tumours in the Key Areas of the Huai Basin'. This project attempted to answer two questions: 'How to define the high incidence of malignant tumours in these areas' and 'Whether water pollution in the Huai River caused residents' malignant tumours'.[18] Later, the project, 'Cancer Prevention and Control in the Huai Basin', was launched, with support from a special national fund. Unfortunately, the results of this project were not made public.

After the media report about Huangmengying, the government also quickly began to improve the water supply in the village. According to my investigations, the government invested in building three deep wells in the county in which the village is located, serving approximately 40,000 people in the surrounding area. Obviously, this was treated as a special case. Since 2000, the Chinese government has put considerable effort into resolving the problem of drinking water in rural areas. The tenth and eleventh Five-year Plans addressed the problem in many areas through the Rural Drinking Water Safety Programme. However, this programme would not have applied to Huangmengying, because it stipulates that, in principle, 'drinking water problems caused by pollution should be resolved by the parties that polluted the water'. In other cases, such as Xiqiao Village and Jiannan Village (see Chapters 5 and 9 of this volume), the author found that although residents requested polluting enterprises to improve the safety of their drinking water, they did not comply.[19]

While it brought some benefits, being reported as a 'cancer village' also caused trouble for the local people. News reports about pollution and cancer were an implicit criticism of the local government, and the negative image they created also affected local investment and economic development. Because of this, in its later work, the local government tried to replace the

18 'China Disease Prevention Control Centre Undertakes an Epidemiological Investigation of Malignant Tumours in the Huai River Basin', 19 August 2005, http://www.chinacdc.cn/zxdt/200508/t20050819_30963.htm

19 Chen and Cheng, 'Risk Perception and Responses to the "Cancer-Pollution" Problem – Empirical Studies in Several "Cancer Villages"'.

narrative about 'cancer villages' with one of 'ecological' (*shengtai*) development. In 2009, I came across a large sign in the village (about 70 centimetres by 1 metre; see Photograph 5) on a wire pole that read, 'Huangmengying Village, a Demonstration Site for the Circular Economy'. In 2011, I noticed that the appearance of the village had greatly improved, and there was a banner in the main street that read, 'Huangmengying Provincial-level Eco-village'. There were also many slogans about environmental protection and ecology on walls along the main streets (See Photograph 6).

Although the TV programme 'River and Village' did describe a real problem of abnormal levels of disease in Huangmengying, it made use of the problem of cancer in the village as a way to raise the issue of water pollution in the Huai Basin more generally, and the focus of the programme was on water pollution from outside the village. It was only after the central government's unexpected intervention that the question of the relationship between pollution and disease in Huangmengying was raised to a new level of national discourse about 'cancer villages'. However, because people were trying to use the situation in the village to point the finger at external pollution, it also became impossible to discuss the original issue of the extent to which there was in fact a clear relationship between pollution and cancer. Meanwhile, institutions such as the Chinese Centre for Disease Control and Prevention, which conducted large government-funded studies in the area, did not officially release their research findings.

Villagers' lifestyles and disease

The original aim of our research in 2009 was to find the sources of pollution behind the high incidence of cancer in Huangmengying. At first, I attempted to pursue my original line of enquiry, in which I assumed that 'pollution was the main cause of cancer', but the reality was much more complex than I had anticipated and my research raised many doubts. After much reflection and discussion, I conducted a second study in 2011. Although it was claimed that pollution from the Huai River was the main factor leading to the high incidence of cancer, our two previous studies and wider consideration of the issues led me to believe that it was related not only to external factors (water pollution) but also to factors within the village, including lifestyle. I reached this conclusion by understanding the village as a whole and the villagers' daily lives. The information below about local lifestyles was generally neglected by the media and the public.

Gender, smoking and lung cancer

When reading the data on people who had died of cancer, I noticed that the sex ratio was unusual. Most of those who died of lung cancer were male. There were 134 names on the list of cancer deaths from 1990 to 2011 and 25 had died of lung cancer, or 18.6% of the total. Nineteen patients were male and six were female (see Table 2.2). On examining their smoking habits, I noticed that fifteen of the nineteen male patients used to smoke and only four did not. Medical research has shown that smoking is related to lung cancer and in Huangmengying most of those who died of lung cancer were male smokers. This shows that residents' lifestyles were also affecting their health.

Table 2.2 Gender and smoking behaviour of lung cancer mortality cases

Sex	*Males*		*Females*	
Number of patients/ percentage	*Number of patients*	*Percentage*	*Number of patients*	*Percentage*
Smokers	15	78.9	0	0
Non-smokers	4	21.1	6	100.0
Total	19	100.0	6	100.0

Hepatitis B and liver cancer

When conducting a study in the Northern Anhui and Eastern Henan areas of the Huai Basin in 2009, I noticed small advertisements for Hepatitis B treatment all over the place and I wondered whether the disease was very prevalent in this area. Dr Wang confirmed that not only were Hepatitis B rates high, but a high proportion of those who died of liver cancer had the condition. From 1990 to 2011, 42 patients died of liver cancer, accounting for 31.1% of all cancer mortality. Most of them suffered from Hepatitis B: 21 of the 42 patients had a clear history of the disease. Only two cancer victims had never had the disease and the status of the remaining patients was unclear. At that time, it was not so easy to test for Hepatitis B and many people did not know whether they had the disease or not. Furthermore, being labelled a Hepatitis B patient would have a bad effect on your marriage prospects and so many families kept it secret. In my 2011 study, I gathered further information about patients who died of liver cancer between 2009 and 2011. Five of them had a clear history of Hepatitis B.

Table 2.3 Gender and history of Hepatitis B patients who died of liver cancer

Sex	*Male*		*Female*	
Number of patients/percentage	*Number of patients*	*Percentage*	*Number of patients*	*Percentage*
Patients with clear history of Hepatitis B patients	19	79.2	2	11.1
Patients without history of Hepatitis B patients	2	8.3	0	0
Patients with unclear status	3	12.5	16	88.9
Total	24	100.0	18	100.0

Dr Wang revealed some information about the families of patients with liver cancer. WXH, who had contracted Hepatitis B, died in 2003 at the age of 38. His brother died in 2008 at the age of 41. Their two uncles also died of liver cancer. SFR died in October 2010 at the age of 48, and his brother died in November 2010 at the age of 49. These two brothers died just one month apart and they both had Hepatitis B. Their father died at the age of about 50 and he also suffered from Hepatitis B. GHR died in 2009 at the age of 61 and his brother also died of liver cancer at the age of 31. His children all have Hepatitis B.

Improvement of the water supply and the decline in digestive tract diseases

According to Dr Wang, after the water supply in Huangmengying was improved, there was a sharp reduction in digestive tract diseases among the villagers. Villagers usually drank unboiled water and before the improvement of the water supply, whenever the weather got warm, there would be an increase in digestive tract diseases. Dr Wang noted the dramatic decrease.

> After the improvement of the water supply, intestinal diseases, such as enteritis, diarrhoea, and chronic colitis, decreased. From May to August, people in this village like to drink unboiled water. During this period of time, intestinal diseases are common. From May to August, I would have over 20 patients with intestinal diseases every day in my clinic… Now I only get one or two. (In August 2011, I asked again how many patients he had per day during a normal stretch of time and he said three or four.) Last time, I made a bet with the chief

> of the county health bureau. When the journalist asked me, I told him I had 30-40 patients every day. The chief of county health bureau heard about that and he came to me and accused me of exaggerating... He said, 'Let me sit in your clinic, and I'll see if you can see 30 patients in a day. I bet that you will get fewer than that'. And so he sat there, but unfortunately for him, I had over 50 patients that day. (Dr Wang interview, 5 April 2009).

I was surprised that Dr Wang said the villagers drank unboiled water. In the past, most people in rural areas used to drink unboiled water; in films produced in the 1950s and 1960s, you see children scooping a ladle of water from a tank and drinking it directly. However, as the economy developed and the quality of water deteriorated, most people in rural areas gave up this habit and started to boil their water before drinking it.

Gradual change in villagers' lifestyles

When I visited Huangmengying in 2011, the village was very different from the way it had been in 2009. One reason for this was that the local government wanted to erase Huangmengying's reputation as a 'cancer villages' and stress 'environmental protection' and 'ecology'. But another reason was that the villagers' lifestyles had changed. Like other economically backward areas, most people in Huangmengying worked outside the village, and through this they increased their incomes and built new houses with better sanitary conditions. Moreover, many residents were influenced by urban lifestyles, and gradually changed some of their unhealthy habits. Dr Wang said that ten years before, people used to eat leftovers such as steamed bread and noodles, but that they no longer did so. Until recently they had also eaten pigs, sheep and chickens that had died of disease, but they also no longer did this and in fact they would not even buy meat that was of poor quality. Overall, the villagers paid more attention to food hygiene, not only in terms of food they produced themselves but also what they bought from the supermarket. They would carefully check whether the food was a 'Three-No' product.[20] Most villagers also kept a piece of land specially to grow grain and vegetables for themselves and would not use pesticides on it even if productivity was low.

20 *Three-No products* usually meant no production date, no quality certificate, and no producer information.

Comparison and discussion

Cancer is a very complicated problem and many issues remain unresolved in medical science; whether pollution causes cancer is even more uncertain. In a situation where science is unclear on many issues, one-sidedly emphasizing one particular factor can lead to biases in perception and practice. In the following section, I use 'internal' and 'external' perspectives as an analytical framework to compare Huangmengying Village with Minamata disease in Japan, with the Tannerstown 'cancer community' in the United States, and with Dongjing Village in China.

Minamata disease was first discovered in Minamata Town in Kumamoto Prefecture, Japan (now Minamata City). The name of the town was later used to name the disease from which its residents suffered. After the Second World War, Japan's economy developed quickly and along with this came pollution. A nitrogen plant in Minamata discharged wastewater containing mercury directly into the bay. Organic mercury was absorbed by fish, shrimp, and shellfish and entered the food system via seafood, damaging the human nervous system and affecting brain function. The causal relationship between pollution and Minamata disease was clear, but there was a huge gap between residents' self-identification as victims and the government's recognition of their claims under the law. In processing applications for compensation, by 1986, the government had confirmed only 1,672 cases and denied 4,999 people who claimed to be Minamata patients.[21] Fully three quarters of residents who claimed to have the disease were not acknowledged by the government. It seems that scientific and legal difficulties were involved, as well as conflicts of interest. Further analysis showed that in situations where there was scientific uncertainty, patients who whose status was not confirmed were more likely to attribute their illness to external causes (pollution), while the government was more likely to regard it as their own problem. These conflicting hypotheses meant that those unconfirmed patients and their supporters (including lawyers) have been protesting since the 1960s. In 1995, the Japanese Prime Minister, Tomiichi Murayama, apologized for the first time, but in the end the government's solution to this issue was driven more by humanitarian concerns than by purely scientific or economic considerations.

Martha Balshem (1993) discussed the way in which cancer was attributed to 'internal' or 'external' factors in the case of Tannerstown in the United States. Noticing the high incidence of cancer in some areas in an earlier study, the

21 Funabashi, 'Minamata Disease and Environmental Governance', pp. 7-25.

government launched a cancer education project, called CAN-DO. Tannerstown, a European-American working- class community in Philadelphia, was a high-cancer-risk area and the government and scientists hoped the project could change residents' lifestyles and lower their risk. But the residents did not believe that these professionals could do anything for them. They argued,

> 'I have a neighbour who eats old food, and she is ninety-three years old…'[22] 'My neighbour's dog just died of cancer. Why did the dog die of cancer? The dog didn't smoke, the dog didn't eat an improper diet, the dog didn't lay out in the sun. So why?'[23]

The residents rejected internal causes and attributed the risk of cancer to environmental pollution. In 1981, the *Philadelphia Daily News* published a series of articles about the 'Cancer Zones' where people were convinced high mortality from cancer was due to industrial pollution.[24] In a series of interviews and focus groups, 93 people said they thought lifestyle factors such as diet and smoking caused cancer, and 185 people – twice as many – selected environmental factors such as pollution and food additives.[25] Clearly the residents had quite different ideas from the designers of the CAN-DO Programme.

In Balshem's study, an ordinary woman, Jennifer, whose husband John had died of pancreatic cancer, bravely challenged the authority of a hospital doctor in an attempt reach an objective understanding of her husband's cancer. The doctor emphasized John's smoking and drinking habits and wrote in his medical record, 'extensive history of alcohol abuse'. Jennifer objected to this and wrote in the margin, 'less than one case of beer per week'. By writing this, she was stressing that her husband had not drunk a lot of beer and that the doctor should not take it for granted that 'alcohol abuse' had caused her husband's cancer. Instead, she highlighted environmental influences on John's health. For example, he had worked in a chemical factory for a long time and lived in a highly contaminated community. She wanted to play down the importance of her husband's lifestyle in his contracting cancer.[26]

In fact, when the CAN-DO Programme was launched, it was premised on the assumption that residents of Tannerstown had bad lifestyles, and that

22 Balshem, *Cancer in the Community: Class and Medical Authority*, p. 59.
23 Ibid., pp. 61-62.
24 Ibid., pp. 23-34.
25 Ibid., p. 71.
26 Ibid., pp. 104-105.

this was the primary cause of the high incidence of cancer. But the residents themselves had different ideas, or at least they wanted to emphasize the role of environmental pollution, a factor that was overlooked by the government and scientists.

Huangmengying was different. In Tannerstown, people within the community stressed external pollution while outsiders focused on internal factors. In Huangmengying, it was outsiders who pointed to external pollution. The movement there was started by environmentalists from outside the village and the story was spread by the national media. The focus of the story was on blaming pollution in the Huai Basin. Although this process drew attention to the problem of cancer, it inevitably introduced some underlying assumptions. In fact, subsequent scientific research found that it was not easy to establish at the village level whether pollution caused cancer or not. The designation of 'cancer village' was influenced by multiple factors but neglect of factors internal to the village led to a partial understanding and response. However, even at times like this, when emotions are running high, these questions still require us to apply a rational, scientific approach in our research.

Further Thoughts on Research from a Village Perspective

The pioneering Chinese sociologists Wu Wenzao and Fei Xiaotong advocated community research in the early days of the last century. They believed that the community (*shequ*) is the basic unit for understanding Chinese society. Wu Wenzao proposed that communities have an observable material foundation, which includes at a minimum the following three elements: (1) people; (2) a place where those people live; and 3) a shared lifestyle or culture. Fei Xiaotong believed that, 'no matter what the reason may be, the village is still the basic unit of Chinese society'.[27]

Village-based community research has certain characteristics. First, it seeks a holistic understanding. Because the location and population of the village is bounded, when the anthropological method of long-term participatory observation is used, it can produce a comprehensive understanding of the community and its interactions with the surrounding environment. Second, community research produces empirical observations, which can be tested to determine whether they are true or false. Third, community research involves multiple dimensions and rich layers, which include the

27 Fei, *From the Soil*. p. 41.

natural geography of the area, farming and land use, sources of economic revenue, and the organization of families, relatives and networks, as well as customs, habits, and religious beliefs.

In particular, it should be noted that the regional features of the place where the community is spatially located have been highlighted in village-based community research, whereas most other sociological studies of social groups are disconnected from a consideration of regional spatial factors. This is particularly important when researching Chinese rural communities because China's territory is vast and there are huge differences among villages. Therefore, it is only possible (although even then it will not necessarily happen) to understand Chinese society through the richness of its villages. As a consequence, the early sociologists placed great emphasis on the geographic and spatial dimensions of the community, As Fei Xiaotong said, 'People's lives are time and region specific and this is the community'.[28] This previous generation of sociologists advocated community research which aimed to reach an understanding of Chinese society at the macro level through the accretion and comparison of empirical knowledge at the micro level. Whether or not this goal can be achieved, this method of research has great significance for understanding rural Chinese society, and this includes exploration of the causes of illness from a social medicine perspective.

In fact, excluding the special groups of the family and the clan, the village is the most important unit of analysis for understanding Chinese rural society. First, as a common unit that transcends individuals and families, it is the basic 'stage' for conducting the public affairs with which sociologists are concerned and where they can conduct empirical observation. Second, China's traditional rural villages have specific geographic environments. The village and its surrounding environment should be able to satisfy the basic needs of the villagers for food and water. The village is thus a comprehensive system which includes the relatively independent factors of its geography and environment, economy, society and culture. At the same time, however, the village is also the most basic unit of society.

If we reflect carefully on the popular term 'cancer village', it is not difficult to discern its sociological significance. Why did this term become so popular? Why do the public and the media not use terms like 'cancer people', 'cancer family', 'cancer clans' or alternatively 'cancer town' or 'cancer county', which have a wider geographic range? Epidemiological research in China usually takes the county as the unit for analysis. For example, in the 1950s and 1960s, a high incidence of oesophageal cancer occurred in Lin County

28 Ibid., p. 85.

in Henan, and liver cancer was widespread in Qidong County, Jiangsu. More recently there was a high prevalence of digestive tract cancer in some counties along the Huai River. In all these cases, the registration of deaths, statistical analysis and research on prevention and control were conducted taking the county as the unit of analysis. In this sense, it would be more appropriate to speak of a 'cancer county'. But in fact, 'cancer village' is the most popular term.[29] The wide popularity of the term 'cancer village' shows that the high incidence of cancer in the village unit has already become a very obvious phenomenon that ordinary people can sense and observe. The focus of the public and the media has therefore already provided us with a very important angle for research. Starting with village, we can reveal the associations between specific geographic, economic, social and cultural factors and human health.

So far, however, medical science research has neglected to study cancer at the village level, and it does not offer a good explanation for the phenomenon of 'cancer villages'. One recently published book, *Environmental Epidemiology, Basics and Practice*,[30] introduced the topic of cancer villages, and this may be the first time it has been included in a medical science textbook. However, the sources of information about the eight 'cancer villages' listed in the book were all media reports from between 2003 and 2010. Three were from the newspapers belonging to the 'Southern' media group including 'Southern Weekend', 'Southern Daily' and 'Southern City', four were from the 'China Business Daily', the 'Hua-Xi Urban News', the 'Changjiang Business News', and the 'County and Town Forum', and another one was from Baidu Encyclopaedia. This shows that 'cancer villages' have still yet to fall within the purview of medical science.

In 2013, a collection of maps, the *Atlas of the Huai River Basin Water Environment: Digestive Cancer Mortality*, provided medical statistical results at the macro level with maps showing the association between pollution in the Huai River Basin and digestive tract tumours at the level of the county. It states,

> When comparing changing trends in patterns of mortality in the Huai River Basin over a 30-year time period, we found that areas with the most severe and longest lasting pollution were also the areas where the mortality rate for digestive tract tumours death rate increased the

29 A search of the Baidu website on 3 February 2015, for 'cancer county', 'cancer rural areas', 'cancer town', and 'cancer village' got approximately 28,000, 4,810, 5,140 and 8,300,000 results, respectively. 'cancer village' produced nearly 300 times the number of search results as 'cancer county'.

30 Zhou and Ye, *Environmental Epidemiology Basics and Practice*, pp. 498-499.

> fastest and it rose at a pace several times faster than the corresponding national mortality rate. The results of spatial analysis reveal a high level of coincidence in the distribution of the most polluted areas and areas with a high occurrence area of digestive tract cancers. This is the most important finding of this collection of maps.[31]

This is the first time scientific statistics have been used to demonstrate the links between pollution in the Huai River Basin and cancer; these assertions are not speculation but supported by data, and they provide important evidence for government policy making. However, this study does not provide more detailed results and the presentation of the incidence of disease at the level of the county unit averages out differences in geographic characteristics within counties, obscuring variation across these smaller units. In fact, it should have been possible to conduct separate analysis of some villages that had serious pollution and a high incidence of disease, and this might have helped to answer some of the pressing questions about the causes. However, because the findings are averaged, some major problems that could have easily been discovered were muted.

Another problem is that when matching disease data with environmental indicators such as Biological Oxygen Demand (BOD), Chemical Oxygen Demand (COD) or ammonia-nitrogen, we cannot tell exactly which pollutants are causing health problems. These indicators can show the situation regarding environmental pollution, but a high content of organic substances and ammonia-nitrogen in water can only prove that it is polluted; it cannot explain exactly what it is in the polluted water that is causing cancer. When it comes to a specific village, such as Huangmengying in Shenqiu County, Henan, we still do not know exactly which or what kinds of chemical substances in the polluted water led to the high incidence of cancer, or whether perhaps water pollution contributed to the abnormal growth of some biological organisms that cause health problems. Analysis of this kind also cannot explain why different villages along the Huai River and its tributaries showed great disparities both in the incidence of cancer and in mortality rates. The authors of the atlas acknowledge that 'the level of evidence provided by this research can answer the question of whether there is a correlation between cancer and environmental pollution, but it cannot answer the question of whether there is an etiological association between the two'.[32]

31 Yang and Zhuang, *Atlas of the Huai River Basin Water Environment: Digestive Cancer Mortality.*

32 Yang and Zhuang, *Huai River Basin Water Environment and Digestive Tract Tumor Death Photo Collection.*

Consequently, I personally believe that medical research, especially studies like that of the high incidence of cancer in the Huai River Basin, cannot be disconnected from analysis of the social groupings and specific environments in which individuals live, and analysis should not use homogenized artificial groupings to replace groups of people which are actually diverse. In keeping with this overall judgement, village-based research can provide new possibilities for examining the causes of cancer. It is an important and meaningful complement to our present understanding of the problem of disease in 'cancer villages' to examine them from the perspective of the behavioural patterns of individuals, relationships within the family, relatives and clans, and through the historical trajectories of the village's geography, economy and society in combination with modern chemistry and biological research methods. For example, the information I found out when chatting with the village doctor about the habits of men and old people in the village alerted me to the relationship between gender, smoking and cancer. Another example is the relationship between changing the water supply and disease, which shows that drinking water habits were related to the high incidence of intestinal disease. People in the area who died of liver cancer had almost all previously had hepatitis C, and the clustering of hepatitis C showed its relationship to social networks. The fact that the water in the village was polluted was also related to its particular geographic location, and to the connections between the villagers' shallow wells, ponds and the irrigation system.

Experience shows that the community is an important battleground in health maintenance and disease prevention. General practitioners (or family doctors) who are based in the community form a natural doctor-patient relationship in the context of the 'familiar society'. General practitioners have a stable set of patients, and their relationship with them does not end when illness is cured but continues and is deepened as they provide a variety of health services over the life course. In the process of doing this they come to understand not just the individual physiology but also the patient's psychology, family and work environments and other factors that influence health. Only by fully understanding these can they give appropriate care and assistance to their patients. Experience shows that the combination of general practitioners and community care is a system that has great added value.

In fact, in the Mao Zedong era, 'barefoot doctors' and the cooperative medical system created a miracle in the history of healthcare in China and the world. The social basis for this was similar to the current system of community health maintenance and disease prevention that is now accepted practice in many developed countries. As everyone knows, barefoot doctors usually graduated from primary or middle school in the village, received only short medical training and were then grouped in practices to provide medical services for 'poor, lower and middle peasants' (*pin xia zhong nong*). At that time, medical skills and standards were very low and economic resources were scant, but the basic health problems of most people in the village could be solved by acupuncture, herbal medicine and limited western drugs and medical equipment.

Many factors contributed to the success of the system, but in terms of understanding the causes of disease, the 'familiar society' in which everyone knew each other very well made it much easier for barefoot doctors to detect health problems. 'Barefoot', in its original meaning, referred to doctors who were from the village and not disconnected from agriculture, but in fact, not being disconnected from agriculture was not the most important thing. What was most meaningful for the rural medical health care system was that barefoot doctors were rooted in the rural environment of the village. Because they came from a local village, these doctors were familiar with the daily lives and production activities of the villagers, and with the economic and social environment in which they lived, including information related to patterns of disease in families and clans that were typical of the location. This made it possible for barefoot doctors to use basic diagnostic skills to effectively and economically diagnose and treat many diseases.

Taking the village as the unit of analysis, and effectively combining sociology and medical science would be helpful for doctors and researchers seeking to understand the causes of disease. The statistical epidemiological analysis mainly focuses on relationships between macro-level variables, while clinical medicine is quite the opposite and is concerned only with individual patients. In fact, in China it is often said to 'see only the disease and not the person' (*jian bing bu jian ren*). But in fact, even seeing the disease and the person is not enough because people live in society, and so the social background of patients is significant for both the diagnosis and the treatment of disease. I have also noticed that, in the actual practice of medicine, some doctors are concerned about family history and background. In this sense, the approach of taking the village in 'cancer villages' as the basic unit of analysis does not overlap with the research subjects of either clinical medicine or epidemiology but can be seen as a new research perspective that can be used to explore the relationships between illness and villagers' lifestyles and the surrounding environment.

Appendix

Table 2.4 List of patients who died of cancer in Huangmengying

The list of people who died in Huangmengying between 1990 and 2009 (according to CCTV's data, between 1990 and 2004 105 people died)

Num.	*Name*	*Gender*	*Age*	*Cancer types*	*Time of death (year, month)*
1	WHT	Male	72	Oesophagus cancer	1990.3
2	WBR	Male	51	Liver cancer (hepatitis B patient)	1990.3
3	WZC	Male	70	Oesophagus cancer	1990.3
4	WBX	Male	70	Liver cancer (confirmed not hepatitis B patient)	1990.3
5	WXC	Male	70	Oesophagus cancer	1990.3
6	WSR	Male	67	Lung cancer (smoked)	1990.3
7	WJQ	Male	66	Oesophagus cancer	1990.4
8	WGS	Female	70	Oesophagus cancer	1990.5
9	WHQ	Male	75	Lung cancer (smoked)	1990.5
10	WBY	Male	72	Liver cancer (not clear whether hepatitis B patient)	1990.7
11	WOS	Female	32	Blood cancer	1990.11
12	WZS	Female	68	Lung cancer	1990.11
13	WS	Female	50	Liver cancer (not clear whether hepatitis B patient)	1991.4
14	WGY	Male	76	Oesophagus cancer	1991.6
15	XZX	Male	65	Liver cancer (hepatitis B patient)	1991.10
16	WS	Female	78	Oesophagus cancer	1991.12
17	TXR	Female	71	Oesophagus cancer	1992.2
18	WSC	Male	28	Lung and kidney cancer (non-smoker)	1992.3
19	WQF	Male	72	Stomach cancer (fire breaks outside)	1992.3
20	WXG	Male	62	Liver cancer (not hepatitis B patient; no family members were hepatitis B patients)	1993.4
21	WWS	Female	56	Colorectal cancer	1993.4
22	WBY	Male	67	Oesophagus cancer	1993.4
23	CHJN	Female	60	Liver cancer (not clear whether hepatitis B patient)	1993.6
24	SJYM	Female	62	Liver cancer (not clear whether hepatitis B patient)	1993.6
25	SQJN	Female	50	Lung cancer (tuberculosis)	1993.9
26	WMR	Female	48	Liver cancer (hepatitis B patient)	1993.11
27	WXB	Male	69	Lung cancer (smoked)	1994.11
28	WBY	Male	45	Bladder cancer (may have suffered from testicular cancer)	1994.2

Num.	Name	Gender	Age	Cancer types	Time of death (year, month)
29	XSS	Female	72	Oesophagus cancer	1994.2
30	WSH	Male	50	Liver cancer (hepatitis B patient)	1994.7
31	WBY	Male	70	Stomach cancer	1995.1
32	SWS	Female	84	Intestinal cancer	1995.2
33	XWJN	Female	78	Stomach cancer	1995.3
34	SSR	Male	67	Liver cancer (hepatitis B patient)	1995.7
35	WHS	Female	58	Liver cancer (not clear whether hepatitis B patient)	1996.1
36	WXC	Male	14	Brain cancer	1996.2
37	SWS	Female	56	Lung cancer (tuberculosis)	1996.3
38	WFC	Male	67	Lung cancer (smoked)	1996.3
39	WCP	Female	46	Intestinal cancer	1996.3
40	WX	Male	36	Brain cancer	1996.4
41	LTZ	Male	68	Oesophagus cancer	1996.4
42	WBS	Male	68	Lung cancer (smoked)	1996.4
43	KCY	Female	63	Liver cancer (not clear whether hepatitis B patient)	1996.5
44	ZGY	Female	62	Stomach cancer	1996.6
45	WXL	Male	60	Oesophagus cancer	1996.6
46	WBG	Male	66	Oesophagus cancer	1996.7
47	WZC	Male	49	Liver cancer (hepatitis B patient; their house was near the western and southern ponds and currently the whole family has migrated away from the village)	1996.7
48	SJYF	Male	66	Liver cancer (hepatitis B patient)	1996.7
49	WQH	Male	38	Lung cancer (non-smoker)	1996.8
50	WHX	Male	72	Lung cancer (smoked)	1996.8
51	WBJ	Male	72	Liver cancer (not clear whether hepatitis B patient)	1996.10
52	GYY	Female	60	Liver cancer (not clear whether hepatitis B patient)	1996.10
53	ZHC	Male	60	Oesophagus cancer	1996.11
54	ZAL	Female	38	Stomach cancer	1996.12
55	WLS	Female	70	Liver cancer (not clear whether hepatitis B patient)	1997.1
56	WZS	Male	68	Oesophagus cancer	1997.2
57	LEW	Male	52	Oesophagus cancer	1997.3
58	WSY	Male	55	Oesophagus cancer	1997.3
59	WKS	Female	70	Skin cancer	1997.4
60	ZXR	Female	25	Stomach cancer	1997.8
61	SXY	Female	37	Oral cancer	1997.8
62	OGR	Female	57	Oesophagus cancer	1997.9
63	WZS	Female	82	Stomach cancer	1997.10

Num.	*Name*	*Gender*	*Age*	*Cancer types*	*Time of death (year, month)*
64	WXL	Male	70	Liver cancer (hepatitis B patient)	1998.3
65	SZS	Female	76	Stomach cancer	1998.3
66	WBW	Male	70	Lung cancer (smoked)	1998.3
67	LXY	Female	44	Liver cancer (not clear whether hepatitis B patient)	1998.4
68	WLS	Female	72	Liver cancer (not clear whether hepatitis B patient)	1998.7
69	WOS	Female	84	Colorectal cancer	1998.8
70	LJX	Male	76	Lung cancer (smoked)	1998.8
71	KYL	Female	63	Liver cancer (not clear whether hepatitis B patient)	1998.9
72	WPS	Female	72	Intestinal cancer	1998.10
73	FXY	Female	52	Oesophagus cancer	1998.12
74	SFC	Male	72	Lung cancer (smoked)	1999.2
75	WXC	Male	62	Oesophagus cancer	1999.3
76	WYS	Female	65	Liver cancer (not clear whether hepatitis B patient)	1999.6
77	WZS	Female	72	Oesophagus cancer	2000.3
78	SAY	Female	50	Liver cancer (not clear whether hepatitis B patient)	2001.3
79	WSH	Male	70	Lung cancer (smoked)	2001.3
80	SFQ	Male	62	Liver cancer (hepatitis B patients)	2001.3
81	XXM	Female	84	Liver cancer (not clear whether hepatitis B patient)	2001.6
82	WDS	Female	74	Colorectal cancer	2001.9
83	ZGZ	Female	52	Oesophagus cancer	2001.10
84	WBJ	Male	62	Stomach cancer	2002.2
85	WLQ	Male	72	Oesophagus cancer	2002.6
86	ZGY	Female	63	Oesophagus cancer	2002.8
87	WBF	Male	60	Brain cancer	2002.9
88	SCS	Female	70	Stomach cancer	2002.10
89	SNS	Female	78	Intestinal cancer	2002.10
90	WXR	Male	77	Oesophagus cancer	2003.2
91	WSY	Male	69	Liver cancer (hepatitis B patient)	2003.3
92	HTE	Male	78	Liver cancer (hepatitis B patient)	2003.3
93	SSF	Male	83	Liver cancer (hepatitis B patient)	2003.3
94	LFZ	Male	80	Oesophagus cancer	2003.4
95	WXH	Male	38	Liver cancer (hepatitis B patient; his brother WCL was a hepatitis B patient and died in March of 2008)	2003.7
96	WGS	Female	79	Liver cancer (not clear whether hepatitis B patient)	2003.7
97	MZJN	Female	68	Liver cancer (not clear whether hepatitis B patient)	2004.2

Num.	*Name*	*Gender*	*Age*	*Cancer types*	*Time of death (year, month)*
98	JSZ	Female	60	Oesophagus cancer	2004.3
99	WXZ	Male	80	Liver cancer (not clear whether hepatitis B patient)	2004.3
100	FWY	Female	60	Liver cancer (not clear whether hepatitis B patient)	2004.3
101	SWS	Female	70	Skin cancer	2004.3
102	WXT	Male	74	Lung cancer (bronchitis, emphysema)	2004.4
103	WXG	Male	78	Liver cancer (hepatitis B patient)	2004.5
104	WXH	Male	62	Liver cancer (hepatitis B patient)	2004.7
105	LCS	Male	72	Stomach cancer	2004.7

The first investigation data (Between 2004 and 2009, 12 people)

Num.	*Name*	*Gender*	*Age*	*Types of Cancer*	*Time of Death*
106	GDL	Male	60	Lung Cancer (smoked)	2004.12
107	SSF	Male	62	Oesophagus cancer	2005.3
108	YXL	Female	55	Leukaemia	2005.4
109	WS	Female	60	Stomach cancer	2005.4
110	XJH	Male	55	Stomach cancer	2005.8
111	WXY	Male	54	Lung Cancer (smoked)	2006.8
112	LYH	Male	60	Lung cancer	2007.3
113	WFL	Male	41	Liver cancer (hepatitis B)	2007.3
114	SFM	Male	61	Lung Cancer (smoked)	2008.11
115	SFZ	Male	63	Oesophagus cancer	2008.12
116	HLS	Female	63	Lung cancer	2008.12
117	HLS	Female	83	Skin cancer	2009.2

The second investigation data (between 2009 and 2011, 17 people)

Num.	*Name*	*Gender*	*Age*	*Types of Cancer*	*Time of Death*
118	WSL	Male	60	Lung Cancer (smoked)	2009.2
119	LGZ	Female	65	Lung cancer	2009.3
120	WZC	Male	65	Oesophagus cancer	2009.8
121	GHR	Female	65	Liver cancer (hepatitis B patient; his brother died of liver cancer aged 31, his son and daughter were also hepatitis B patients)	2009.11
122	SQF	Male	70	Oesophagus cancer	2009.3
123	WCL	Male	43	Liver cancer (hepatitis B patient; his brother WFL and two uncles were all liver cancer patients)	2009.10

Num.	Name	Gender	Age	Types of Cancer	Time of Death
124	XS	Female	71	Emphysema, lung cancer (non-smoker)	2010.3
125	HLS	Female	92	Skin cancer	2010.5
126	GEY	Female	39	Pancreas cancer	2010.8
127	WXL	Female	75	Oesophagus cancer	2010.6
128	SFR	Male	48	Liver cancer (hepatitis B patient; 128 and 129 were brothers)	2010.10
129	SAS	Male	49	Liver cancer (hepatitis B patient; their father was a hepatitis B patient, died of liver cancer aged 50)	2010.11
130	SZ	Male	55	Lung Cancer (smoked)	2010.10
131	JJF	Male	78	Oesophagus cancer	2011.8
132	LKY	Male	75	Cholangiocarcinoma cancer	2011.7
133	LYLM	Female	92	Bladder cancer	2011.6
134	WBE	Male	54	Liver cancer (hepatitis B patient; his son was also a hepatitis B patient)	2011.3

Works Cited

Balshem, Martha. *Cancer in the Community: Class and Medical Authority*, Washington and London: Smithsonian Institution Press, 1993.

CCTV (China Central Television), *River and Village*, 10 August 2004. http://www.cctv.com/news/china/20040810/102281.shtml. Accessed 6 June 2012.

Chen, Ajiang. *Secondary Anxiety: The Social Interpretation of Water Pollution in Tai Lake Basin*. Beijing: China Social Sciences Press, 2010.

Chen, Ajiang and Cheng Pengli. 'Risk Perception and Responses to the "Cancer-Pollution" Problem – Empirical Studies in Several "Cancer Villages".' *Xuehai* 3 (2011): 30-41.

Chen, Ajiang, Luo Yajuan, and Chen Tao. 'The Paths and Consequences of "Pollution First, Clean Up Later" – A Case Study of Environmental Protection and the Maoji Leather Industry.' *Journal of Guangxi University for Nationalities (Philosophy and Social Science Edition)*, 3 (2011): 55-60.

Development Office of the State Council Leading Group on Poverty Alleviation, *List of Key Counties for National Poverty Work*. 2012. http://www.cpad.gov.cn/art/2012/3/19/art_50_23706.html. Accessed July 5, 2013

Fei, Xiaotong. *From the Soil.* Berkeley, CA: University of California Press, 1992.

Funabashi, Harutoshi. 'Minamata Disease and Environmental Governance.' *International Journal of Japanese Sociology*, 15, no. 1 (November 2006): 7-25.

Lin, Mei. 'Game Theory Analysis, Relationships and Patterns: An example of the Prevention of Pollution in the Huai River Basin.' In *China Environmental Sociology – a Discipline Under Construction*, ed. by Hong Dayong, 342-343. Beijing: Social Science Publication Press, 2007.

Lu, Jianbang. 'The Orientation of Cancer Prevention and Control in China as Viewed from the Development of Esophageal Cancer Scene.' *China Cancer*, 2010 (issue 1): 11-14.

National Development Reform Committee. *National Rural Drinking Water Safety Project – The 11th Five-Year Plan*. 2006.

Song, Guojun (ed.). *China Huai River Basin Water Environment Protection Policy Evaluation*. Beijing: China Renmin University Press, 2007.

Wang, Supin. *Liu Xing Bing Xue* (*Epidemiology*). Beijing: Peking Union Medical College Press, 2009.

Yang, Gonghuan, and Zhuang Dafang. *Atlas of the Huai River Basin Water Environment: Digestive Cancer Mortality*. Dordrecht: Springer. 2014.

—. *Huai River Basin Water Environment and Digestive Tract Tumor Death Photo Collection*. China Map Publication Press, 2013.

Zheng, Liandi. *The Water Conservancy Encyclopedia of China: The History of Water Conservancy*. Beijing: China Water Power Press, 2004.

Zhou, Yikai, and Ye Linxiang. *Environmental Epidemiology Basics and Practice*, People's Health Publication Press, 2013.

3 A Subei 'Cancer Village'[1]

Luo Yajuan

Abstract

From 2000, local governments actively recruited polluting industries to move into rural areas of Northern Jiangsu (*Subei*). This chapter analyses the case of Dongjing Village. It describes the actual situation regarding industrial pollution, the interests and choices of the local government, the villagers' resistance, the actions and limitations of the media, and the disconnect between environmental law and the actual handling of conflicts over pollution. The research found that the attitude and actions of the government showed a tendency towards post hoc crisis management. But the problem of industrial pollution is difficult to solve, and the appearance of 'cancer villages' during this period was related to factors including the cadre evaluation system, the structure of power and interest relations, and the lack of independence of administrative agencies.

Keywords: chemical industry, high incidence of cancer, catch-up modernization, environment and health, media, environmental law

Introduction

The term 'Subei' has referred to different geographical areas in different historical periods. Right after the establishment of the People's Republic of China (PRC), 'Subei' referred to an administrative area under the 'Subei People's Administrative Office' which was established in Taizhou City on 21 April 1949. The Subei Office administered five cities – Taizhou, Yangzhou,

1 Part of this article was previously published in *Xuehai*, Issue 2, 2010, under the title 'Environmental Protest against Rural Industrial Pollution – Case Study of Dongjing Village'. The names of towns and village, enterprises, and people have been changed.

Chen Ajiang, Cheng Pengli, and Luo Yajuan, *Chinese "Cancer Villages": Rural Development, Environmental Change and Public Health.* Taylor & Francis Group 2020
DOI: 10.5117/9789089647221_CH03

Yancheng, Huaiyin and Nantong – as administrative divisions. In January 1950, the Subei Administrative Office decided to combine the Yangzhou and Taizhou administrative divisions,and named them the Taizhou Administrative Area. The Subei Administrative Office was then relocated from Taizhou to Yangzhou. In April of the same year, it was renamed the Jiangsu Province Subei Administrative Office. In November 1952, the Subei administrative area, the Sunan administrative area, and the city of Nanjing joined to form Jiangsu Province.[2] In early 1984, based on research in the small towns of Jiangsu, Fei Xiaotong suggested replacing the traditional two-way division into Sunan and Subei with a three-way division of Jiangsu Province into three areas: Sunan, Suzhong and Subei.[3] Sunnan includes the three cities of Suzhou, Wuxi and Changzhou, and the eastern and southern parts of Nantong city; Suzhong includes part of Yangzhou City along the Yangtze River, Zhenjiang, Nanjing, and the western and northern parts of Nantong City; and Subei includes Xuzhou, Lianyungang, Yancheng, Huaiyin, and part of Yangzhou City. According to the current division of the Jiangsu Administrative areas, Subei has 40 counties (cities and districts), including five prefecture level cities: Xuzhou, Lianyungang, Suqian, Huai'an, and Yancheng. According to the 2010 'Jiangsu Statistical Yearbook', the land area of Subei is 54,357 square kilometres, and at the end of 2009 the population was 33.35 million.[4]

In more recent times, the word 'Subei' has acquired richer economic, social, and cultural connotations. First, 'Subei' is an economically depressed area in comparison with 'Sunan' and 'Suzhong'. Before the mid-Qing dynasty, the prosperity of the Subei region was comparable to Jiangnan, and there was not yet a geographical concept of 'Subei'. The impoverishment of Subei is related to the declining importance of the Beijing-Hangzhou Grand Canal and the changing of the path of the Yellow River in the mid-nineteenth century. Starting in the 1920s, the original function of the Grand Canal was gradually replaced by ocean shipping, causing Subei to lose its position as an important centre of transportation and commerce. As a result, the government neglected the maintenance of the channels and sluices connected to the Grand Canal, and this caused an unprecedented number of disastrous floods in the region. The change in the path of the Yellow River in the 1850s exacerbated the problem.

2 Jiangsu local blog network: http://www.jsdfz.com/Article.asp?articleid=388

3 Fei, 'Small Town, Subei First Exploration', pp. 76-77.

4 Statistics from: Jiangsu County Statistics webpage, 'Jiangsu Statistical Yearbook – 2010', electronic version, http://www.jssb.gov.cn/jstj/tjsj/tjnj/.

After the founding of the PRC, the region benefited from a series of water conservancy projects, including the Subei General Irrigation Canal (See Photograph 7). Floods were brought under control, and agricultural production improved substantially. However, the lack of local industry and the region's distance from the economic hubs of Shanghai and other major cities meant that there was still a huge gap between the economic situation of Subei and that of Sunan and Suzhong. Catching up with Sunan has therefore been an ongoing development goal for Subei municipal and county governments.

Second, the disasters, famine and poverty that have characterized the region in modern times have led 'Subei' to be gradually reconstructed as a derogatory term representing inferior cultural status. Subei people who migrated to Shanghai and other places due to war, floods, or famine often undertook the harshest jobs with the lowest pay and the worst reputation, such as pulling rickshaws. In the mid-nineteenth century in Shanghai and other parts of Jiangnan, 'Jiangbei' (primarily 'Subei'), and 'Jiangbei People' gradually became synonyms for low status. To call someone a 'Jiangbei Pig' meant that regardless of whether this person actually originated from Jiangbei or not, they were poor, clumsy, dirty and vulgar.[5] This kind of cultural bias still exists today.

The use of the word 'Subei' in this article is based on the economic homogeneity of the Subei region. After the founding of the PRC, and especially since economic reform, the area has been labelled a place of poverty and backwardness; pursuing modernization and development but under the secondary anxiety of being compared to the Sunan region.[6] This meant that there was a particularly strong internal demand for industrialization and this affected the way in which all levels of the local government thought about the trade-offs between basic human needs and environmental protection. At the early stage of industrialization, although the dangers of industrial pollution were understood, the preference was initially to lean first towards sacrificing the ecosystem to achieve economic growth, bringing in chemical, printing, dyeing, and metal plating and other heavily polluting industries from Shanghai, Sunan, Eastern Zhejiang and other developed areas, with a relatively low environmental barrier to entry. Polluting enterprises sprouted up all over the Subei countryside, and it was pretty much true that every chemical factory polluted at least one village.

5 Emily Honig has fully documented this in her work *Subei People in Shanghai, 1850-1980*.

6 The concept of 'Secondary Anxiety' comes from the work of Chen Ajiang, *Secondary Anxiety: A Social Interpretation of Water Pollution in Taihu*.

Dongjing village is one such case. This research documents the situation regarding polluting enterprises in the village, the interests and behaviour of the local government, the harm caused to villagers and their struggle, the actions and limitations of the media, and the effectiveness of environmental law in adjudicating environmental disputes. It presents a comprehensive picture of the political, economic, and socio-cultural backdrop to a Subei 'cancer village'. The purpose of this research, and the title of this chapter, 'A Subei Cancer Village', is to provide a warning to other areas, especially those are even less developed than Subei. From a national macroeconomic standpoint, the economy of Subei falls in the upper middle band in terms of development. This means that half the regions in China are facing the heavy task of industrialization that Subei has already grappled with, and they are at a crucial turning point in terms of their choice of the path of development.

In terms of methodology, this research is exceptional in its use of a longitudinal approach. Since the research team's fieldwork first began in 2007, we have continued to pay close attention to Dongjing. Team members conducted a total of 4 field trips to the village. In August 2007, Luo Yajuan and Zhou Yin went there for the first time to investigate the process of the enterprise's establishment in the village, the onset of pollution and harm, and the villagers' resistance. In January 2008, Luo Yajuan, Li Caihong, and Song Liangguang returned to Dongjing and conducted further research on the state of pollution from the enterprise, the suffering of the villagers and their struggles. In March 2009, team members went to Dongjing for the third time to follow up on ecological and social changes after the relocation of the polluting enterprise. Chen Ajiang, Luo Yajuan, Chen Tao, Geng Yanhu, and Shen Shuzheng participated in this investigation. In October 2011, the team returned to Dongjing for the fourth time, to research the situation regarding the recovery of the ecosystem, and the health and livelihoods of the villagers, as well as the attitude and opinions of the local government towards industrial pollution in Dongjing several years after the incidents. Luo Yajuan and Zhou Yan participated in this investigation. In addition to official fieldwork investigations, the research team was in frequent contact with Dongjing villagers. Constant observation over a long period of time helped us understand the causes and consequences of the polluting enterprise entering the village, its long term impacts, and its relocation. It also enabled us to observe the situation regarding cancer rates after the relocation of the polluting enterprise. This information is crucial for a comprehensive understanding of the village.

The Chemical Plant Arrives in the Village

Our survey found that in 2007 the total population of Dongjing Village[7] was 2,115 and per capita arable land was 1.1-1.2 *mu*.[8] Administratively, Dongjing belongs to Liuji Town, Funing County, Jiangsu Province. The east of Funing County faces the Yellow Sea; and the west lies beside the old path of the Yellow River. The terrain is high in the west and low in the east, and it is criss-crossed by canals. The Sheyang River runs through the middle of the county, and the Subei General Irrigation Canal runs through the northwest. The county has a dense network of rivers and a multitude of lakes. It has a northern subtropical monsoon climate, suitable for growing various crops, such as rice, wheat, corn, soybeans, canola, and cotton. The transport in this area is very convenient, with a coastal highway on the east side of the county, the Yanxu Highway on the west, National Route 204 in the middle, and Provincial Highways 329 and 231 crossing the whole county. The total area of the county is 1,439 square kilometres. Of this, arable land covers 88,751 hectares and water covers 35,157 hectares. There are more than fourteen towns in the jurisdiction, a Funing County Development Zone, which has provincial level status, a Dong Yi Economic Zone and an Ao Yang Industrial Park, including a total of 338 village committees, 2,477 village groups, and sixteen residents' committees. There are 355,000 registered households in the county, with a total population of 1.094 million, of whom 378,200 are non-agricultural residents.[9]

If Subei can be said to be the economic 'low-land' of Jiangsu Province, then Funing County is located at the lowest point. Prior to 1949, Funing County's key characteristics could be summarized with the words 'floods' and 'poverty'. Due to the county's location near the sea and next to the river, it often suffered from floods caused by the Yellow and Huai Rivers, as well as by coastal tides. Every year from late spring to late autumn, under the influence of hurricanes and typhoons, the tide rises, often resulting in disasters. Floods caused poverty, and ordinary people who relied on the land for food had no choice but to leave their homes and resort to begging. Because there were a lot of beggars in Funing and the two neighbouring

7 Dongjing Village refers to the population that lives around Julong Chemical Plant, consisting of several natural villages (eleven villager groups). It is a victim community of Julong Chemical Plant. For the geographical relationships between Dongjing Village, its land and rivers, and Julong Chemical Plant, see Figure 3.1.

8 A *mu* is approximately 0.0667 hectares.

9 Statistics from: Funing County Government webpage http://www.funing.gov.cn/, statistical data from 2009. Accessed 10 August 2010.

counties (which were commonly called the 'Three Northern Counties' by people in Yancheng City), when people mentioned them, they immediately identified these counties as poor.

After the founding of the PRC, a series of irrigation projects helped free the people of Subei from the suffering caused by floods and constant hunger. In 1951-1952, the Subei General Irrigation Canal was dug. It starts from the Gaoliang River at Hongze in the west of the county, and runs to the Biandan Harbour at Sheyang, spanning 160 kilometres and giving the Huai River access to the sea. The initial function of the canal was to alleviate drought and provide irrigation in the dry season, and to prevent against floods in the flood season, adjusting water levels during normal times and facilitating transportation. After the flooding was dealt with, agricultural production recovered, and the lives of local residents stabilized. However, due to the history of poverty and lack of local industries, Funing County's economy was still relatively underdeveloped, and it faced enormous pressure to catch up. As recently as 2006, Funing stood out even among the 16 counties in Jiangsu designated as priorities for poverty relief and development work, with a total of 60 'poverty villages' in the county.[10]

The pressure to relieve its impoverished situation and catch up with the surrounding area in Subei and the Sunan region translated into a strong motivation to attract investment. One paragraph on the Funing County government website summarizes the efforts made to attract investment in recent years and their results.

> In recent years, our county persistently used an open policy to promote development, increasing vitality through foreign investment, using projects to strengthen the primary sector, using entrepreneurship to enrich the people, and attracting investment as the biggest chance for development [...] the whole county established 10 professional Investment Promotion Bureaus [...] with an emphasis on going to Sunan to identify industries that were being eliminated there and attract them to transfer to Funing, engaging Taiwanese investment from Guangdong, and attracting Korean investment from Shandong. During the period of the 'Tenth Five-year Plan', the county has implemented a total of 250 industrial projects over 10 million yuan. Of these, there are 205 projects with an individual worth of 10-50 million yuan, 25 projects worth 50-100 million yuan, and

10 Statistics from: the State Council Poverty Alleviation Leadership Team Official Webpage, http://www.cpad.gov.cn/data/2006/0303/article_312.htm. Accessed 8 May 2008.

> 20 projects worth over 100 million yuan. In 2006, the county implemented 146 projects with an individual value of over 10 million yuan, of which 11 exceeded 100 million yuan.[11]

It is against this regional backdrop that Julong Chemical was invited into Dongjing Village, as a 'God of Wealth'. In early 2000, Julong moved into the Liuji Town Industrial Park situated in the middle of Dongjing as part of the town government's effort to attract businesses and investment. Although it was called an 'Industrial Park', the site did not actually qualify as such. It consisted only of Julong Chemical, two very small textile factories, and an abandoned paper factory, situated on a strip of farmland by the side of the Yanhuai Road. The Liuji Town government adopted a series of preferential policies to attract Julong Chemical to set up the facility there: waiving the cost of land, and partially reducing taxes for the first two years of operation. Julong Chemical occupied 25 *mu* of land, had almost 10 million yuan in fixed assets, and an annual profit of around 1.3-1.5 million yuan. (See Photograph 8).

When Julong Chemical first arrived, the villagers of Dongjing did not understand the actual production processes it used. Officials from the town and village, as well as representatives of the chemical plant, told the villagers that the factory would absorb local surplus labour and bring prosperity to the local economy. At the same time, they promised that the plant produced flu capsules and other chemical products in daily use, and that it used a water recycling technology and would not spill a single drop outside the plant or cause any pollution to water sources or the air in the surrounding area. As one villager recalled,

> Julong Chemical Plant came in March 2000, as the result of the town government's policy of attracting investment. At the time, we villagers all said that chemical plants cause pollution, and that we do not agree to the plant being sited in our village. Officials from the town and village as well as people from the chemical plant said they only produce cold capsules, vanishing cream, and other similar daily chemical products, and there won't be any pollution, so we agreed. After that the villagers helped the chemical plant build the foundation, and when that was done, they started constructing the factory building. The construction lasted right up until September, and then production began. (Villager Shu Qichang Interview 1, 28 August 2007).

11 Statistics from: http://www.funing.gov.cn/zsyz/20070323/164929.html. Accessed 10 May 2008.

Figure 3.1 The geographical relationship between the Julong Chemical Plant and the Dongjing Village communities, land and river

The promises made by the town and village officials and the chemical plant representative reassured Dongjing villagers who had been worried about pollution and other potential hazards from the Julong plant. A factory opening right outside their door, not having to migrate for work, the addition of another stable source of employment and income, and with no risks involved – this was the kind of industrial development the villagers had been longing for. With dreams of making their fortune, they helped the chemical company construct the foundation for the plant, to make it easier for them to become operational and contribute to the prosperity of the village as soon as possible. Julong Chemical thus settled smoothly into Dongjing. The geographical relationship between the plant and Dongjing Village communities, the surrounding land and the river is shown in Figure 3.1.

'Settling Scores' with the Chemical Plant

The First Experience of Pollution

In September 2000, Julong Chemical began production, and as soon as it did, the peculiar smells emanating from the plant caused the villagers to doubt the previous promises that there would be 'no pollution'. Recalling the state of emissions from the plant at the time, Shu Qichang said:

> We were suspicious of what they were producing. The smoke coming out of the chimneys smelled very bad and very strong. Once the smell of the material they used got out we couldn't stand it. The northwest wind blew it to the southeast, and the northeast wind blew it to the southwest. When the wind blew southwest it came to our place, and that smell was there day and night, so people in our village began to question what on earth they were producing in this chemical plant! (Shu Qichang Interview 1, August 2007).

Right after that, there were cases of fish and shrimp dying in the river and the flesh of the fish had a strange odour that also made the villagers feel the 'force' of the waste water emitted from the Julong Chemical Plant. (For the situation of sewage entering into rivers and farmland, see Photographs 9 and 10; for a typical spot for local water use, see Photograph 11). Duan Qirong recalled:

> After the chemical plant came, our river water started turning red. At first, whenever the chemical plant released water, a lot of fish would float to the surface. Some of them were very big fish that had grown over many years, and after floating for a while, they all died. At first, some people took those fish home to eat, or sold them in the market, but first they had to be kept in clear water for several months to remove the smell [...] and usually after eating them once and noticing the strong odour, you wouldn't dare to eat them again. So people would give those fish to their cats and dogs, but even they wouldn't eat them. Some people dried them in the sun, but even then there was still a strong smell and cats wouldn't eat it. (Duan Qirong Interview 4, January 2008).

After the harvest that autumn, the villagers discovered that the pollutants emitted by Julong Chemical seemed to have got into their rice, sweet potatoes, and other crops. Dongjing villagers realized that Julong Chemical's promise of 'not producing pollution' was a guise to deceive them and obtain approval for the plant.

> Later, when we were threshing the rice, we all thought it was strange that our rice had the smell of pesticides. Even the rice without husks had this smell, and this had never happened in the past. When we boiled the new rice,

> the whole room was full of the smell of the chemical plant, and we dared not eat it. We took the dried sweet potatoes and glutinous rice dumplings we made to relatives who lived in other places, but when they ate them they said they smelled of pesticides and nobody dared to eat them. Later we figured that the chemical plant began production in September, which was just around the time when the rice started flowering, and the smell followed the pollen into the grains. (Shu Qichang Interview 1, August 2007).

Initially, the villagers tried to transfer these impacts away from themselves. For example, in 2000, some of them gave the newly harvested rice to the government to get tax credits or to sell, and they bought rice in from elsewhere to eat themselves. Different villagers felt differently towards the risks from pollution. Some firmly refused to eat their strange-smelling rice, while others chose to put up with the smell and eat it anyway. These differences in attitudes and choices were related to villagers' knowledge as well as their economic situation.

A Trip to 'Settle Scores'

Our story comes to the winter of 2000. There is not much farm work in this season and the villagers are idle most of the time. This was when some villagers found out that the things Julong Chemical Plant was producing were not actually flu pills and vanishing cream and other daily chemical products; the products it made were actually being supplied to pesticide factories. The abundance of free time in the winter season gave villagers the chance to gather together to chat, and the matter of Julong Chemical Plant's deception gradually spread through the community. The villagers protested that they had been cheated.

In spring of 2001, they began ploughing the ground, fertilizing the canola and wheat, and planting vegetables. The temperature rose again, and a strong southeast wind began to blow. Villagers in Groups 1 and 2, which were situated downwind of Julong Chemical (see Figure 3.1) were plagued by waste gas emissions from the factory on a daily basis. At the same time, they discovered that the vegetables they planted had a strong odour that was difficult to wash off. Villagers usually do not use pesticides on the vegetables they consume themselves, and so they concluded that the chemical smell could only have been caused by pollution from the Julong Plant. After eating these vegetables some villagers felt unwell. This, and the bad air quality, sparked their discontent, and they began to discuss how they could 'settle scores' with Julong Chemical.

> The people in Groups 1 and 2 discussed going to the chemical plant to settle scores. At that time, everyone's dissatisfaction with the factory had been stirred up and we people over here (Group 6) belong to the same village as they do, so we went too. We all surrounded the chemical plant and wouldn't let them continue production.
> The workers in the plant were mostly brought over by the company from Jianhu County [Author's Note: the factory director of Julong Chemical is from Jianhu County, Yancheng City], and few of them are from here. Young people from Jianhu County said they would beat us up, but we were not afraid of them, there were more of us! In the end they did not dare to raise their fists. Although we disrupted it for some time, the factory did not agree to stop production, so some people from our side started to dismantle the factory chimney, smashing the windows and doors, and blocking up the pipe that supplied water to the plant. (Shu Qichang Interview 1, 28 August 2007).

The Julong Chemical Plant had a production license, and so the director naturally refused to negotiate with the villagers, much less compromise. The two parties reached a stalemate, and Julong Chemical's representative made a phone call to the Liuji Town government. The villagers believed that the chemical plant had been the first to break its promise, disrupting villagers' agricultural production and daily lives. They felt that they had done nothing wrong and that they were fully justified in wanting to talk with the town government. In fact, the outcome shows that the villagers overestimated the possibility for them to dialogue with the local government, and that they underestimated the risks they faced. The town government protected Julong Chemical and gave a demonstration of power by 'killing the chickens to scare the monkeys' (making an example of some of the villagers to intimidate the others).

> Government people said it was illegal for us to smash the windows and doors, and to dismantle the chimney. They said that the chemical plant was in production, and we villagers were sabotaging it, disrupting production and public security. A total of eleven people were detained [...], and [...] the others didn't dare to cause trouble anymore. The longest period of detention was 38 days and the shortest was three days. The town government only released those people after they were made to promise not to cause disturbances at the chemical plant in the future and threatened them by saying that if they didn't agree they would be relocated to another place and detained. The masses are very timid, and after this nobody dared cause any more trouble, so everything calmed down. (Shu Qichang Interview 1, August 2007).

Aside from those in Groups 1 and 2, a few other leading villagers led the other residents in the protest against the factory. I believe there are two reasons for this. First, suffering from the same environmental hazards made the villagers develop a common sense of identity as victims of pollution standing together in opposition to the chemical plant. Second, it was possible to mobilize the villagers because they felt their actions were justified. They did not participate in the action casually; every villager made a rational decision about whether to take part, and the rationale for their participation was very clear – the factory had breached the promise it had made when it was set up and so the villagers were justified in seeking to settle the score. It is exactly this sense of shared identity and the rational justification for their decisions that united villagers who had initially been scattered like loose sand.

Why did the villagers first approach the chemical plant, and not the town government? When they decided who to protest against, they made an unconscious distinction between 'friend and foe'. The chemical plant entered the village by means of deception, and its pollution brought harm to the villagers, making it an absolute 'enemy'. It was the town government that had brought the chemical plant into the village, and the officials had also deceived the villagers about whether the chemical plant would emit pollutants, Despite the fact that they are supposed to play the role of 'mother and father' to the common people, it was still questionable whether the local officials would act in the cause of justice. It was not clear whether they were really enemies or friends. In actual fact, this first protest served as a test case for future actions in terms of determining whether the local government was an 'enemy' or a 'friend'. In all future protests, the villagers regarded the town government as an 'enemy' and went over its head to report their problems to higher levels of government.

Why did the villagers' 'justifiable' protest result in failure? On the surface, it was a contest between the chemical plant and the community, but in fact, it was a contest between the local government and the villagers. To understand why the protest failed, we therefore need to consider it from the perspective of both these parties.

First, why was the local government so overly eager to attract investment? He Xuefeng believes that the enthusiasm for attracting investment in China at the time was not because local governments were so eager to increase revenue, but rather the result of external pressure. Under the strong top-down administrative system, higher levels of government assigned investment targets to lower levels, and used these as the main criterion for assessing officials' performance.[12] I agree with He Xuefeng that in seeking to attract investment, local government officials were primarily concerned about their own political performance evaluation. Key officials at the town level typically serve a three-year term of office. To compete with each other in attracting investment, they offered preferential policies that would appeal to investors. One beneficial tax policy that was commonly offered was 'three years' exemption and two years 50% discount', under which the enterprise had no obligation to contribute to the local government's financial revenue for the first three years. This meant that local government representatives were not generally attracting investment with a view to increasing revenue, which would only come after their term of office had ended. But the 'annual amount of contracted outside investment' was used by higher levels of government as an important indicator to evaluate the performance of officials at lower levels. The weight given to inward investment in performance evaluations was usually much larger than the weight given to social development, ecological development and other indicators. Under this performance evaluation mechanism, local government officials pursuing career development were under very heavy pressure to attract investment.

Second, the fact that local governments could invite polluting enterprises in and protect them was also due to the way the relationship between the local government and the villagers is structured. The villagers were in a much weaker position in terms of their right to express their interests. The local government represents political power and controls the discourse. Because of this, even though the chemical plant had destroyed the villagers' lives and agricultural production, the town government turned this on its head, defined the villagers' protest as 'destroying production', and detained some of them. In this sense, the result of this protest was decided long before it even began.

Villagers Outside the Protest: Onlookers

Not all villagers joined the resistance to the factory. Some villagers observed from a distance. Duan Qirong, who later mobilized the villagers to engage in a lawsuit, is an example. Duan is an elderly man, over 60 years of age at the time, whose rich life experiences made him very cautious when dealing with such matters. He did not join the protests at first, partly because he

12 He, *New Rural China – Research Notes on Rural Society During the Transition Period*, pp. 40-44.

had a certain degree of legal awareness, but also because he predicted the form the retaliation by the local government might take.

> The Liuji town government helped Julong Chemical to deceive us. Their saying that Julong produced cold pills and vanishing cream means they knew what it was actually producing, and they knew it would cause pollution. This also means that they were already taking the side of the chemical plant. Given that they're taking the side of the chemical plant, they will not help us ordinary people. If we go and cause trouble, and engage in illegal behaviour, then they will have a justification for not helping us anymore. (Duan Qirong Interview 2, 29 August 2007).

> We can only try our best not to do things that are illegal; otherwise it will only have disadvantageous results for us. Last time they went to the chemical plant to fight, I did not join. The resistance I'm leading against the plant will not tolerate anyone doing anything illegal, or we will be put in the wrong. (Duan Qirong Interview 4, 27 January 2008).

Duan Qirong made an accurate judgment about the results of the protest, but he did not stop the villagers from going, and instead assumed the position of an onlooker. Why was this? First, he wanted to confirm whether the position of the town government was really as he had guessed; and second, he still harboured the hope that the government would side with the villagers and help them solve the pollution problem. He hoped that the villagers' actions would be effective, since from his perspective, the chemical plant did indeed cause pollution, and the town government should also stand by its initial promise. In the end, the results of the villagers' actions were in accord with his expectations, and confirmed his judgement, but this also made him very disappointed.

'Two-Faced People'

Aside from onlookers like Duan Qirong, two other types of people did not join the resistance in Dongjing. The first was village officials and the second was villagers who were working at Julong Chemical. Village officials in China typically act on the one hand as representatives of the villagers, and on the other as the executives of government power in the countryside, or representatives of the town government. When the requirements of these two roles conflict, and the situation is not handled properly, village officials can fall into the awkward trap of offending both their constituency and their superiors. In Dongjing, village officials had to obey the town government officials, and help stop the villagers causing trouble, but at the same time they could not openly offend the villagers. The villagers were very clear on this point too.

> The village officials did not join the lawsuit, but they did not dare to act against us either. Because the town government supports the chemical plant, village officials often collect information on us for the town government. But it's also not good for village officials to 'stiff' [*lao*] us. [Author's Note: 'Lao', local dialect, meaning to make a relationship rigid/stiff]. If village officials openly opposed us, the villagers would oppose them. (Duan Qirong Interview 4, 27 January 2008).

In Dongjing, the positions of Village Party Secretary and Village Head (Chairman of the Village Committee) are held by the same person. When I arrived in the village in 2008 for the second round of fieldwork, the Village Secretary at the time happened to have died from oesophageal cancer not long before, and so I did not have the chance to interview him. The interesting thing was that the villagers all believed that the village secretary dying of oesophageal cancer was retribution for bringing the chemical plant to the village. From their point of view, he was in the circle of people with an interest in the chemical plant and had benefited from it financially. In an interview conducted in October of 2011, when the case had become less sensitive, village officials told us their real thoughts about their conflicted situation at the time:

> Pollution from chemical plants is harmful to human health. Nobody wants a chemical plant right outside their door. But because the higher level of government was pressuring us, we had to work to change the attitude of the local people and prevent the villagers from petitioning and bringing law suits. The main issue was that the higher levels of government needed the revenue from the chemical plant. The fact that the chemical plant is now gone is mostly down to the villagers starting lawsuits. (Village Head Wang Interview, 9 October 2011).

The villagers working at Julong Chemical comprised another special group. If Dongjing villagers were generally weak, then the villagers working at the chemical plant were the weakest of the weak. More physically robust, knowledgeable, and strong-minded villagers typically worked outside the village. The ones who worked at Julong Chemical were generally less advantaged in terms of their physical attributes and cultural capital, they were from relatively poorer families, and their families often had special circumstances that prevented them from working far away from home. Working at the chemical plant, although it only paid 600 or 700 yuan a month, provided much needed income for them. They were therefore both beneficiaries of the pollution and the victims of it. As such, they experienced very strong contradictory feelings and adopted the position of onlookers.

From the Government to the Media

The villagers from Groups 1 and 2 who led the resistance against the chemical plant were scared off by the government. Inside, they were even more dissatisfied with the chemical plant and with the town government, but out of concern for their personal safety, none of them wanted to lead the resistance anymore. The action of the town government in detaining some villagers had had the cautionary effect of 'killing the chicken to scare the monkeys'. Most of the villagers now understood which way the wind was blowing, and chose to suffer the pollution caused by the chemical plant in silence. Although resentment boiled in their hearts, no one was willing to stand up in protest. What Mancur Lloyd Olson called 'The Logic of Collective Action' was quietly playing out in the village.[13] If one considers Dongjing as a large group, every villager, when deciding whether to contribute to the common good of the village environment, first assessed the effort they needed to exert, the possible risks, and the possible individual gains. Any rational villager, after evaluating the situation, would decide not to take any risks to contribute to the common good, and no one would be willing to use their own 'lead barrel' to stop the 'flood'.[14] The result of this failure

13 Olson describes in *The Logic of Collective Action* the plight of collective action. The main idea is that a large collective or potential collective will not organize to take collective action, even though as a collective they have a reason to do so. The reason is that in a large collective, each individual is insignificant; his individual efforts will not provide significant effects for his collective, and no matter whether he contributes to the good of the collective, he will enjoy gains brought about by others. Therefore in situations regarding public goods where there is a lack of incentive, individuals in a large collective will be unwilling to contribute to the common good.

14 The analogy of lead barrels stopping the flood comes from Olson's *The Logic of Collective Action*. It states that when collectively facing a flood, each individual with a lead barrel will feel that their own strength is insignificant, and will be unwilling to face the risk of being washed away to stop the flood. But the result of everyone's logical choice is that there will be no-one to stop the flood. This is a classic example of a collective action tragedy.

to take collective action was a flood of pollution, spreading unrestrained, and in the end, all the villagers suffered even worse environmental hazards.

When things quietened down after the villagers' attempt to settle scores in the summer of 2001, Julong Chemical changed its visible waste water pipes for a 'hidden channel'. This exited from the back wall of the plant and followed the outer wall along a small irrigation ditch that ran between the fields of Group 9 and then flowed towards the river in front of the village that is part of Group 4. (Figure 3.1 shows the path along which the waste water flowed). At the same time, the factory entered a period of feverishly high production, operating day and night. It produced 2000 tons of p-chlorophenol annually and also started production of 500 tons of 2,4-dichlorophenol and 50 tons of phloroglucinol, as well as simultaneously producing o-chlorophenol, triclosan, and Diclofenac. The peak of the production period was also the peak of the pollution and it caused a series of hazards.

Pollution Damages: Declining Returns for Agricultural Sideline Activities

From 2001, villagers found that their rice and other crops were stunted and producing significantly less. They therefore had to increase the amount of fertilizer they used to boost the growth and quantity of crops, increasing their production costs.

> This chemical plant's waste water and gas both have very strong odours and the waste water is dumped straight into the river in front of my home [Author's Note: Villager Liu Yongxin lives in Village Group 4, the river he mentions is marked on Figure 3.1 as river 3]. A few years ago there were no living creatures left in the water. Now the water level is higher and the water also disperses faster, so it's not so smelly anymore. But our farmland was very seriously affected. The rice in other areas only needs one bag of urea per *mu* to grow very well; over here even one and half bags is not enough! (Villager Liu Yongxin Interview 1, August 2007).

Villager Duan Qirong did some quick calculations, which clearly demonstrate how the increase in the cost of agricultural production, reduced output, and lower prices lead to a decrease in villagers' income.

> Our village's per capita arable land is around 1.1-1.2 *mu*. Every year we plant one crop of rice and one of wheat. Before the pollution, the rice output per *mu* was around 500 kg, with a sale price of 90 cents/500g.

> After the farmlands were polluted, rice production fell to 400-425kg, and because our rice has a medicinal smell, we have to sell it for 10 cents/500g less than other people, or around 80 cents/ 500kg [...] Now it's a little better, but when the pollution was serious the rice fields typically needed 75kg of urea per *mu*, or 15kg more, which cost 30 yuan. And even with more fertilizer the rice still didn't grow well. The cost of pesticide also increased by 10-15 yuan per *mu*. (Duan Qirong Interview 4, 27 January 2008).

From these observations, we can calculate that the price of rice decreased by 260-305 yuan per *mu* after the pollution occurred. Dongjing's arable land per capita was 1.1-1.2 *mu*, so on the assumption that there are three people per household, each family had a total of 3.3-3.6 *mu* of arable land and the decrease in income from rice alone was 858-1098 yuan. At first, the villagers' rice was sold at relatively low prices, but later nobody dared to buy rice produced in Dongjing at all because of the strong smell. None of the rice merchants in the town wanted it. After this, a small percentage of villagers slowly changed their habits, and started planting saplings instead.

Agricultural sideline activities do not bring much income, but they are crucial for supplementing families' livelihoods. In Dongjing, almost every household was accustomed to keep pigs, sheep, chickens, and ducks. This was usually done on a small scale, of three or four pigs, five or six sheep, and ten to twenty chickens and ducks. Keeping chickens and ducks did not bring in much profit but for an ordinary farming household, the income from selling eggs could be of vital importance when there were temporary shortfalls in the family's finances or extra expenses, for example buying pesticides, buying food for guests, or snacks for children. Keeping sheep and pigs brought somewhat higher profits; in the years when the price was high, five or six sheep could bring a profit of thousands of yuan, and three or four pigs could bring in as much as a family earned from planting wheat or canola.

A village is an ecosystem, in which all living things are inter-connected. After 2001, the pollution caused by Julong Chemical worsened daily and its effects on the livestock and fowl of farming households also deepened. Many chickens and ducks died and the surviving birds mostly stopped producing eggs. The few that did continue to lay produced eggs that had a strange smell. Sheep and pigs were frequently ill, sows often died when pregnant, or gave birth to still-born piglets. Income from these sideline agricultural activities fell dramatically.

Pollution Damages: Villagers' Experience of High Cancer Rates

From their everyday life experience, Dongjing villagers believed that increased pollution had led to a higher rate of cancer and other diseases. Duan Qirong kept statistics of the cancer incidence and death rate in the village between 2001 and 2005[15] (the peak of Julong Chemical's production and the peak of pollution). Over these five years, there was a total of 82 cancer patients, of whom 64 died and 18 survived. The average incidence of cancer in Dongjing over these five years was therefore 775.4/100,000, and the average annual mortality rate was 605.2/100,000. The 2001 cancer mortality rate in rural China was 112.57/100,000, so Dongjing's annual average mortality rate was 5.37 times higher.

We can see from Tables 3.1 and 3.2 that of 82 cases of cancer, the majority were cancers of the digestive system. The total number of digestive system cancer cases was 54, or 65.5% of the total (40 cases of oesophageal cancer, 1 case of throat cancer, 1 case of cardiac carcinoma, 11 cases of stomach cancer, and 1 case of colorectal cancer). Of these, oesophageal cancer had the highest onset rate, accounting for 48.2% of the total. Looking only at individual types of diseases, oesophageal cancer (40 cases), stomach cancer (11 cases), and liver cancer (14 cases) all had a high incidence.

As can be seen from Tables 3.1 and 3.3, the annual mortality rate for oesophageal cancer in Dongjing was abnormally high, at 19.9 times the annual national rural mortality rate for this kind of cancer. Liver cancer followed, at 5.1 times the annual national rural mortality rate, and then stomach cancer at 3.6 times. The lung cancer mortality rate was also 2.2 times the national rural rate. Although, as I discuss below, the area had a higher than average rate of cancer since statistics were first recorded, with similar patterns of disease, the levels for some cancers had also risen considerably and it is likely that industrial pollution played at least some part in this.

We discovered in follow up research that: (1) Funing County, Yancheng City, where Dongjing Village is located, has had a relatively high cancer mortality rate for a long time, especially for oesophageal, stomach, liver and other digestive system cancers. In 1971-1975 the annual average cancer mortality rate in Funing County was 147.06/100,000 and in Yancheng City it was 138.23/100,000, and in 1973-1977 the national annual average cancer mortality rate was 73.99/100,000, or only round half of this. In 1971 to 1975, the

15 Statistical data on Dongjing Village's cancer rate and mortality rate was collected by Duan Qirong, I verified the data, and deleted a few cases from the beginning of 2000 (before Julong started production).

Table 3.1 Dongjing Village 2001-2005 record of the deceased

Serial Number	*Patient's Family Name*	*Type of Cancer*	*Time of Death*	*Serial Number*	*Patient's Family Name*	*Type of Cancer*	*Time of Death*
1	Yan	Stomach Cancer	2004.10	33	Liu	Liver Cancer	2005
2	Lu	Liver Cancer	2003.06	34	Liu	Oesophageal Cancer	2001
3	Zhou	Oesophageal Cancer	2005.07	35	Bian	Oesophageal Cancer	2001
4	Lu	Oesophageal Cancer	2003.04	36	Duan	Lymphoma	2003
5	Li	Stomach cancer	2005.10	37	Wang	Stomach cancer	2002
6	Chen	Stomach cancer	2003.08	38	Liu	Oesophageal cancer	2002
7	Qiu	Oesophageal cancer	2005.03	39	Liu	Lung cancer	2002
8	Wang	Brain tumour	2005.03	40	Yang	Oesophageal cancer	2002
9	Song	Stomach cancer	2003.03	41	Dai	Stomach cancer	2002
10	Yan	Stomach cancer	2003.03	42	Dai	Oesophageal cancer	2005
11	Yang	Oesophageal cancer	2004.02	43	Wang	Liver cancer	2002
12	He	Oesophageal cancer	2004	44	Duan	Liver cancer	2002
13	Yang	Oesophageal cancer	2001	45	Gu	Oesophageal cancer	2001
14	Tao	Oesophageal cancer	2003	46	Gu	Oesophageal cancer	2002
15	Wang	Liver cancer	2001	47	Zhang	Liver cancer	2003
16	Chen	Oesophageal cancer	2001	48	Xu	Oesophageal cancer	2002
17	Chen	Oesophageal cancer	2003	49	Gu	Lung cancer	2003
18	Gu	Stomach cancer	2003	50	Chen	Oesophageal cancer	2004
19	Qiu	Oesophageal cancer	2005	51	Chen	Oesophageal cancer	2005
20	Chen	Oesophageal cancer	2001	52	Gu	Cardiac carcinoma	2001
21	Yang	Oesophageal cancer	2004	53	Gu	Liver cancer	2004
22	Gu	Oesophageal cancer	2002	54	Li	Oesophageal cancer	2005
23	Chen	Oesophageal cancer	2001	55	Li	Oesophageal cancer	2005

Serial Number	*Patient's Family Name*	*Type of Cancer*	*Time of Death*	*Serial Number*	*Patient's Family Name*	*Type of Cancer*	*Time of Death*
24	Tan	Liver cancer	2002	56	Li	Lung cancer	2005
25	Li	Liver cancer	2003	57	Li	Liver cancer	2005
26	Sun	Oesophageal cancer and lung cancer	2002	58	Li	Lymphoma	2001
27	Zhou	Liver cancer	2002	59	You	Oesophageal cancer	2005
28	Yao	Endometrial cancer	2003	60	Ma	Endometrial cancer	2004
29	Yang	Liver cancer	2004	61	Dai	Oesophageal cancer	2003
30	Jin	Oesophageal cancer	2002	62	Chen	Lung cancer	2001-2005
31	Liu	Oesophageal cancer	2001	63	Shu	Liver cancer	2001-2005
32	Sun	Oesophageal cancer	2002	64	Chang	Liver cancer	2001-2005

Note: In the above table, patients 62-64 contracted cancer or died between 2001-2005, but the exact time of death is unclear.

Table 3.2 Dongjing Village 2001-2005 record of cancer survivors

Serial Number	*Patient's Family name*	*Type of cancer*	*Onset time*	*Serial Number*	*Patient's Family name*	*Type of cancer*	*Onset time*
1	Lu	Stomach cancer	2001.06	10	Shu	Oesophageal cancer	2001
2	Liu	Lung cancer	2005.03	11	Ji	Oesophageal cancer	2002
3	Wang	Lung cancer	2005.07	12	Zhou	Oesophageal cancer	2001
4	Chen	Oesophageal cancer	2003.02	13	Qiu	Oesophageal cancer	2004
5	Chen	Colorectal cancer	2001	14	Zhang	Stomach cancer	2001-2005
6	Zhang	Oesophageal cancer	2005	15	Dai	Stomach cancer	2001-2005
7	Tong	Endometrial cancer	2003	16	Xue	Oesophageal cancer	2001-2005
8	Liu	Lung cancer	2004	177	Xue	Oesophageal cancer	2001-2005
9	Guo	Throat cancer	2005	18	Sun	Lung cancer	2001-2005

Note: In the above table, patients 14-18 contracted cancer between 2001-2005, but the exact onset time is unclear.

Table 3.3 Annual rural cancer mortality rate and Dongjing cancer mortality rate

Type of cancer	*National rural cancer mortality rate*		*Dongjing Village*	
	Year 1991	*Year 2000*	*Number of deceased (persons)*	*Mortality rate (1/100,000)*
Total	101.39	112.57	64	605.2
Nasopharyngeal	1.70	1.78	–	–
Oesophagus	16.32	15.24	32	302.6
Stomach	22.55	20.94	8	75.7
Colorectal	5.14	6.04	–	–
Liver	22.25	26.06	14	132.4
Lung	14.29	21.11	5	47.3
Breast	1.51	2.11	–	–
Blood	3.31	3.18	–	–
Cervix	2.31	1.91	–	–

Note: National rural cancer mortality statistics are drawn from: 'Report on China's Cancer Control Strategy', *China Tumor* (2002), 5th edition, p. 253

average mortality rates for oesophageal, stomach, and liver cancer in Funing County were 62.55/100,000, 41.95/100,000, and 15.23/100,000, respectively, or 4.2, 2.7, and 1.5 times the national annual average mortality rate. In Yancheng City, the annual mortality rates for oesophageal, stomach and liver cancer were 55.78/100,000, 32.92/10,000 and 16.71/100,00, or 3.7 times, 2.1 times and 1.7 times the national average mortality rate, respectively. Because of this, Yancheng City and Funing County were listed as having high rates of cancer overall, and particularly high rates of oesophageal and stomach cancer. (2) The cancer mortality situation in Dongjing in 2001-2005 showed some similar patterns to the data for 1971-1975. The three digestive system cancers – oesophageal, stomach and liver cancer – all had a relatively high mortality rate, and oesophageal cancer had the highest rate. This type of cancer accounted for 50% of all cancer deaths in Dongjing in 2001-2005; in 1971-1975, oesophageal cancer accounted for 42.33% and 40.36% of all cancer deaths in Funing and Yancheng. This historical data shows that the high rate of oesophageal cancer in Subei preceded industrialization and therefore must also have other contributing factors. Other members of our team pursued this question in research in Lianshui County which is discussed in the next chapter.

Families with cancer victims firmly attributed the illness of their relatives to pollution from Julong Chemical. The farmhouses of Tong Damei and Shen Yuzhenin in Village Group 9 were around a kilometre away from the plant, and their fields were right outside the factory, separated

Table 3.4 Cancer mortality rates in Dongjing, Funing and Yancheng

	Dongjing Village (2001-2005)		*Funing County (1971-1975)*		*Yancheng City (1971-1975)*		*National Average (1973-1977)*	
	Mortality rate (1/10,000)	*Percentage (%)*	*Mortality rate (1/10,000)*	*Percentage (%)*	*Mortality rate (1/10,000)*	*Percentage (%)*	*Mortality rate (1/10,000)*	*Percentage (%)*
Total	605.2	100	147.06	100	138.23	100	73.99	100
Oesophageal Cancer	302.6	50.00	62.55	42.53	55.78	40.36	14.95	22.34
Stomach Cancer	75.7	12.50	41.95	28.51	32.92	23.82	15.41	23.03
Liver Cancer	132.4	21.9	15.23	10.40	16.71	12.09	10.09	15.08
Other	84.5	15.6	27.33	18.56	32.82	23.73	33.54	39.55

Note: The statistical data on the cancer mortality rate in Funing County, Yancheng City area in 1971-1975 comes from the Jiangsu Province Yancheng City area Revolutionary Committee Health Bureau Development and Reform Commission Technology Branch, 'Compilation of Research Materials on the Cause of Death from Cancer' (1976). Data on the national cancer mortality rate in 1973-1977 comes from the Health Bureau Cancer Prevention and Cure Research Office Chief Editor, *China Malignant Tumor Death Research* (Beijing: People's Medical Publishing House, 1980).

only by a wall. In 2008, when she agreed to be interviewed, Tong Damei was 59 years old. She had contracted endometrial cancer in 2003 and was cured in 2004, after spending more than 20,000 yuan in medical fees. Her husband died of oesophageal cancer and lung cancer in 2005, after spending more than 50,000 yuan on treatment. The couple's combined medical expenses ran to more than 80,000 yuan and the high debt they incurred pushed them into poverty. Tong Damei's opinion about the cause of their cancer was:

> In the past we were both very healthy. But after the chemical plant came, the grain, vegetables, and water we consumed all had problems, and slowly we got sick. The broad beans we planted had an odour even after being peeled, and could not be eaten. The soy beans we planted smelled strange, and could not be sold. This year my mother-in-law died of old age, and we don't even have money for the funeral. In three years, two people died in my family. The cancer over here is very serious; it's a famous cancer village. (Tong Damei Interview 1, January 2008).

Three members of Shen Yuzhen's family contracted cancer, one after another. Shen Yuzhen herself contracted colorectal cancer in 2001. She was already cured when we interviewed her but she had spent more than 30,000 yuan in medical fees. Shen's husband died in 2004 of oesophageal cancer, after spending more than 10,000 on treatment. Her brother-in-law[16] contracted oesophageal cancer in 2007, and died because of the family's overwhelming debt and lack of money for medical treatment.

> Our family's rice field is right outside the walls of the chemical plant, and the impact was very severe. Our family always ate the grain we planted ourselves, and didn't buy grain from other places, because the family doesn't have the means to buy or exchange grain. We also did not go to the chemical plant. It's useless even if we do go. We aren't educated, can't read, and speaking out has no effect, it's no use for ordinary people like us. Now there is only me at home, my son went to Wuxi to work. If they don't go out to work and earn money, we won't have any way to pay back the debt. They also have another consideration for going out and working, that is my grandson is so small, they can't stay over here and eat the grain here [...] My eyes are not good anymore; I often can't see clearly, they all say I cried too much. My husband died, now my brother-in-law also died, and there is only me left at home. My son once went to the People's Court of Yancheng to sue the chemical plant. He saw our family being affected by the factory, and that several people got cancer, and he felt it was unjust. But he went to sue them by himself without any evidence and it did not work. (Shen Yuzhen Interview 1, January 2008).

From Shen Yuzhen's account, we can sense the enormous financial and emotional damage caused to peasant families by illness and death. At the same time, we can see from her narrative that in making arrangements for family members, helping young people avoid pollution was a consideration; her son and his wife took the grandson with them to work outside the village, avoiding the effects of pollution.

Some other villagers also left Dongjing out of concern over the impact of pollution from the chemical plant on their health. Before the factory arrived in the village, young people who went outside the village for work usually left their children in the care of the older generation. After the chemical plant came, some villagers encouraged these young people to take their children with them. Other young people who originally worked on farms in the village also went away to work one after another because they were unwilling to tolerate the pollution. Dongjing Village elementary school is situated directly west of Julong Chemical, only about 500 metres away. The school originally had more than 200

16 Unmarried, does not have an independent family, lives together with the family of little brother.

students, but after the pollution problem started, the villagers gradually began transferring their children to elementary schools in other villages or in the town. Due to the decrease in the number of students, Dongjing Village school was closed by the Town Office of Culture and Education. Aside from this, some villagers with financial means and relatively strong earning capacity bought houses in the county town. But villagers with average economic means could not afford to relocate, and typically had to rely on the land to earn a living, so they had no choice but to stay in the village and suffer the pollution. From this perspective, the lower the means of a peasant family, the less able they were to deal with the hazard of pollution, and the more harm they faced, which often resulted in their falling into even harsher poverty. On this level, it can be said that environmental hazards contributed to the reproduction of poverty.

Seeking Help from the Government

Intensifying environmental pollution and associated damages provoked some villagers' slumbering resentment, but the majority still remembered the nightmare of their previous episode of resistance, and concerned for their personal safety, they were unwilling to use their 'lead buckets' to stop the 'flood'. After the previous effort, they generally thought that returning to the chemical plant to protest would not be effective because the town government benefited from the plant. However, they believed that the higher level of government was well-intentioned, and so between 2002 and 2004, Shu Qichang and ten other Communist Party members, along with Duan Qirong and twenty other villagers, petitioned continually by writing letters of complaint to higher levels of government. They first wrote to and visited the County Environmental Protection Bureau (County EPB), making clear the situation with pollution from Julong Chemical and the harm they had suffered, The villagers hoped the County EPB could regulate the factory, but they never received a reply to any of their letters or phone calls. They waited in vain for someone from higher up to come and investigate the situation and, meanwhile, the chemical plant continued production at the same feverish rate as before. The villagers continued to petition, to the Municipal Environmental Protection Bureau, the Provincial Agricultural and Forestry Department, and the Provincial Environmental Protection Office, but the response they hoped for never came.

> So a few villagers wrote letters and phoned the County EPB, and then the Municipal Environmental Protection Bureau, but it was useless [...] the car of the Environmental Protection Bureau drove away alongside cars owned by the chemical plant, and the officials did not ask villagers

> what the real situation was! After that people came from the Provincial Environmental Protection Office too, but as before, they ate and drank with people from the chemical plant, and when we called again the Environmental Protection Office people said they have been here to look, and there were no problems! If they don't come to learn about the situation from the masses, the problem will never be solved. (Shu Qichang Interview 1, August 2007).

The villagers placed too much hope in the higher levels of government and environmental protection departments because they believed they were independent of local interests and specialized in inspecting problems. They did not realize that higher level government agencies do not usually handle petition letters directly, but instead instruct lower level departments to deal with the situation themselves. And the lower level of government has all kinds of 'flexible methods' for dealing with this type of issue.

> Before the environmental protection bureau people came to the chemical plant for inspections, they even called the factory in advance. So the chemical plant prepared everything, they stopped production, and they cleaned the workshop and welcomed the environmental protection bureau people in front of the plant. (Shu Qichang Interview 1, August 2007).

After a few years of petitioning in vain, Duan Qirong's feeling was that 'the lower levels of government don't deal with things in accordance with the regulations, and officials protect each other'.

Stirring up Public Opinion

In mid-March 2005, a pollution incident occurred that led the villagers to turn to the media for help: Julong Chemical's gutter became clogged. At the time, the factory's phloroglucinol line had been in production for two months, operating day and night and discharging a large quantity of waste water. On 29 March, the waste water quickly filled the gutter and overflowed into villager Liu Yongda's fish pond. Several thousand kilograms of fish soon floated belly up on the surface of the water. Indignant, Liu Yongda sought compensation from the chemical plant. The two parties reached an agreement that the factory would compensate him 2,800 yuan, and he would not pursue the issue.

On the same day, this incident became known to Liu Yongda's cousin Duan Qirong. Duan believed this incident was strong evidence that the waste water from the chemical plant was dangerous, so he called the County EPB, demanding that it carry out its duty, and punish the plant for its illegal emissions. The EPB did not respond at all, so Duan and a few other villagers decided to go to the media to expose this incident; in Duan Qirong's own words they wanted to 'to stir up public opinion and drive the factory out'. Duan frequently watched a programme called 'Legal Life' on Yancheng TV, and so he called the programme and on 1 April a team came to Dongjing and made a programme, which they promised the villagers they would broadcast within a week. But the villagers waited for an entire month and did not see it on the air. When they called the production team to inquire, the response they received filled them with indignation:

> When we called again, they said the video material was taken away by the County Government, and they could do nothing about it... We heard rumours afterwards that our town government and the chemical plant gave the TV production team 30,000 yuan and that settled it. (Shu Qichang Interview 1, 28 August 2007).

The villagers were extremely frustrated. They believed the County Government and County EPB were both protecting the chemical plant. Afterwards, they used the incident of Liu Yongda's fish pond as evidence of the hazards from the waste water released by the chemical plant, and put pressure on the County EPB, writing a letter demanding that it fulfil its regulatory duties. The letter was sent at the end of April 2005, but the villagers never received a response,[17] so they decided to petition the State Environmental Protection Agency (which later became the Ministry of Environmental Protection).

On 23 October 2005, the villagers' petition to the State Environmental Protection Agency was transferred to the office of the China Economic Times newspaper.[18] A reporter from the paper, Pan Xiaoming, took the

17 In fact, the villagers received a response to their letter from the Environmental Protection Bureau one year later on 20 May 2006. The reason why the EPB provided a response a year later is addressed below in the context of the villagers' lawsuit.

18 The original text of Pan's article read, 'On October 23, 2005 a petition letter arrived at the China Economic Times publishing house after many setbacks'. The precise nature of the 'setbacks' was not indicated. ('Jiangsu Province Yancheng City Cancer Village: Pollution Causes Hundreds to Die of Cancer in the Previous 3 Years'.) This article can no longer be found on the China Economic Times webpage, but many websites re-published it at the time, which proves that the article originated from the China Economic Times. Examples of re-published

villagers' petition letter, a letter of introduction from the State Environmental Protection Agency and his reporter's badge,[19] and went to Dongjing. The villagers felt that the saviour they had longed for had finally arrived. Shu Qichang recalls:

> The reporter was a young man, and he said he was making a private investigation on behalf of the State Environmental Protection Agency. When he arrived by bus, he got off just in front of Old Duan's (Duan Qirong) pesticide shop. He got out his reporter's badge, business card, and letter of introduction from the State Environmental Protection Agency and showed them to Duan, telling him that he had come from Beijing to make a private investigation. He also took out the materials we mailed to the environmental protection bureau, and Duan Qirong recognized them. They were the materials we had sent.
>
> So Duan Qirong took the reporter home and made some phone calls and brought a few of us over, and we told Pan Xiaoming the whole experience of the chemical plant from beginning to end. In the afternoon we took him to the chemical plant to take photos. The factory was taken by surprise, and had not made any preparations. Inside the plant the odour was so strong you couldn't bear it, and the young reporter had to walk around holding his nose. He couldn't stand the smell, and came out very quickly. After we left, the chemical plant called the town government to send some people to keep an eye on us.
>
> The young reporter planned to go to the County EPB and the Municipal EPB the next day, and then return to Beijing When we got to the County EPB, the bureau head wouldn't even come out to see us, and they told us the directors had all gone to Shenzhen for a study session. The officials wouldn't come out, and there was nothing we could do. The young reporter just sat there in the director's office and because he had an introduction letter from the State Environmental Protection Agency, the County EPB people didn't dare to chase us away. So they called the Liuji Town Government and the chemical plant, and in less than half an hour, they arrived. The EPB director also appeared at this point, saying that he had returned from his study session in Shenzhen. They told us it was past ten o'clock, and they would arrange for us to have lunch – the reporter would be invited by the EPB, and we would be invited by the chemical

websites: Yahoo, http://news.cn.yahoo.com/051101/1049/2fucd_4.html; Tengxun, http://news.qq.com/a/20051102/001720_2.htm.

19 As related by villagers.

> plant. We did not agree. We said we wouldn't eat the chemical plant's food, and the young reporter also refused to eat with them. We went downstairs and found a taxi, and the reporter said we'd go to Yancheng City and have lunch there.
>
> Cars from the government and the chemical plant followed right behind us, and the taxi wasn't as good a model as theirs so we couldn't lose them. We took the taxi to the No.204 state highway to wait for a bus going into Yancheng City, but they still followed us. The Liuji Town officials and the chemical plant people knew we were going to the EPB in Yancheng City, so they went there themselves [...] When the reporter realized what was going on, he decided to have a colleague interview the Yancheng EPB at a later date and that he would go straight back to Beijing. So we had lunch with the young man and then took him to the train station.
>
> Afterwards, officials from the Town Government went to the newspaper office [in Beijing] to find the young reporter, but he had already told security not to let anyone from Funing County in, so they were stopped outside the door. See how determined the people of our town government were – they followed him all the way to Beijing, and waited there for two days. (Shu Qichang Interview 1, August 2007).

Pan Xiaoming conducted a lot of inquiries with the villagers, the town government, and the County EPB, and on 3 November he published an article in the China Economic Times about how Julong Chemical had polluted the environment in Dongjing and this had affected the villagers' health.[20] The County Government Office made a response refuting the report and I later obtained a copy of this from Zhang Dafu, who helped the villagers with the lawsuit. The title was 'Briefing on the Inaccurate Report Published in the China Economic Times about Pollution from Julong Chemical Co. in Yancheng City Causing a 'Cancer Village'. In the briefing, the County Government attributed the 'inaccuracies' in the article to the fact that 'the reporter only listened to the inaccurate reports of Duan Qirong and his two acquaintances, and other masses incited by them, and did not conduct any further research.' In the period immediately after this, many journalists went to Dongjing to interview villagers, but it was not long before the media's enthusiasm waned. The villagers realized that stirring up public opinion was not as effective as they imagined; nobody came to solve their problem

20 Statistics from Jiangsu Province Yancheng City Cancer Village: Pollution Causes Hundreds to Die of Cancer in the Recent 3 Years: http://news.sohu.com/20051102/n227363025.shtml. Accessed 21 January 2020.

and the pressure of media exposure did not lead the government to regulate the chemical plant.

It can be seen from this episode of resistance that no matter whether the villagers made phone calls or sent letters to the government, they always skipped the township level of government. This demonstrates that they had already learned the lesson of their resistance in 2001, and understood where the township government stood. But why was sending letters and making phone calls to higher levels of government ineffective? Why did the EPB not fulfil its regulatory duties? And why did turning to the media to generate pressure from public opinion also fail to produce any practical results? The reason why directly petitioning higher levels of government did not work was that, under the prevailing administrative arrangements, higher levels of government did not handle such issues themselves as the villagers hoped, but instead instructed the lower level government or the EPB to deal with the problem themselves. Therefore, although the villagers' appeals took several turns around the bureaucracy, they were still eventually transferred back to the local government. In the end, whether the problem raised by the villagers would be solved still depended on the attitude of the local government.

Why did the local government not act? After 2003, the chemical plant began to make direct contributions to the local government's revenue; this is the main reason why the local government protected the plant. According to an introduction by Liuji Township Party Secretary Zhang Kelei, 'as a key investment project, the chemical plant has made a great contribution to the town government [...] Its tax contributions have increased from 1.3 million yuan to 2 million yuan this year (2005), the annual fiscal revenue of Liuji Township is over 3 million yuan'.[21] According to calculations provided by Zhang Kelei, the tax provided by Julong Chemical in 2004 was two thirds of the total fiscal revenue of Liuji Township. So the government's attitude was that 'the chemical plant has made a big contribution to the township government'. At the same time, starting in 2006, agricultural taxes[22] were abolished, and so tax revenue from industry became more important to the local government.

The EPB's inaction was mainly due to the fact that it was not an independent entity in terms of either its personnel or its finances. In fact, the EPB may

21 Statistics from: 'Jiangsu Province Yancheng City Cancer Village: Pollution Causes Hundreds to Die of Cancer in the Previous 3 Years': http://news.qq.com/a/20051102/001720_2.htm.

22 Agricultural taxes were a kind of tax imposed on all work units and individuals engaged in agricultural production. From 1 January 2006, all agricultural taxes were abolished. This reduced the burden on farmers, but it meant a decrease in revenue for local government.

also have been a direct beneficiary of the chemical plant. As Chen Ajiang has pointed out, the law permits environmental protection departments to have legal income – County or Municipal EPBs can fine polluting industries within their jurisdiction, and the money becomes their own income. In this situation, the 'environmental carrying capacity' of an area could become an important resource for enriching government agencies.[23]

The media constructs the content of 'news', but it is not the only actor to do so. As we can see from the case of Dongjing, local media is not independent, and their construction of the news is restricted by the local government. National media like the China Economic Times are independent of local government, and so the paper was able to expose the hazards caused by pollution from the Julong Chemical and generate pressure from public opinion. But, in the end, even this kind pressure could not succeed in driving out the chemical plant as the villagers hoped. As we saw, the local government had its own tactics for dealing with the pressure of public opinion caused by the media.

Lawsuit

The pollution in Dongjing continued to worsen and environmental hazards increased. From the end of 2005 to the beginning of 2006, during the slack period for agriculture, Shu Qichang, Duan Qirong, Chu Yuhua, and Bian Xiaoren gathered together and discussed what they might do to drive Julong Chemical out of the village and seek justice. They thought of taking a legal approach. Their first thought was that the national law states that the EPB is in charge of issues relating to the environment, and responsible for supervising and monitoring the state of pollution. They therefore decided to start a lawsuit and sue the EPB for failing to act. In the words of Shu Qichang, they would be 'suing the Environmental Protection Bureau for failing in its responsibilities'.

They sought out Zhang Dafu, a lawyer in a firm in a neighbouring town, to be their agent. The 55-year-old Zhang used to be a farmer. He had a middle-school education, with a self-taught knowledge of the law and a great interest in legal research. Zhang loved to stand up for people who were suffering injustice. He often helped people to file lawsuits, and had gained a reputation in the surrounding towns. But even after much consultation, and a large amount of preparatory work, when Zhang Dafu agreed to take

23 Chen, 'Analysis of Stakeholders in Incidents of Water Pollution', pp. 169-175.

up the villager's case he did not have much confidence that they would win, because he had not previously dealt with a lawsuit in which 'common people sue officials'.

On 2 March 2006, Duan Qirong, Shu Qichang, Chu Yuhua, and Bian Xiaoren filed an Administrative Litigation Suit with the County People's Court.[24] The four villagers requested that the County EPB conduct supervision and regulation of the environment around the Julong Chemical Plant, and fulfil its legal responsibilities of environmental supervision and management. The outcome of this lawsuit, in the words of the villagers, was that 'in the end it was not clear who had won and who had lost'. On the one hand, the court recognized the legal effect of the twelve 'Supervision Reports' provided by the EPB and ruled that the agency had fulfilled its responsibilities. In this regard, the villagers believed themselves to have lost. However, the court also recognized that after the incident with Liu Yongda's fish pond, the EPB did not respond to the villagers' request that it fulfil its legal responsibilities to conduct environmental supervision, and it demanded that the County EPB provide a response within 30 days. On this count, the villagers saw a victory within their loss, and this was why, on 20 May 2006, the County EPB finally responded to the villagers' letter of the previous year demanding that it fulfil its responsibilities to supervise and regulate the environment. The letter read,

> To adapt to structural adjustment, and the need to concentrate industries, at the request of the County Party Committee and County Government leaders, on 23 April 2006, a complete relocation of Julong Chemical Co., Ltd, will be implemented. The plant must halt production of phenol-chloride products that release traces of aromatic hydrocarbons before 31 May 2006 and its production of phloroglucinol, which does not cause pollution, must be relocated completely to the County Ecological Chemical Park by the end of 2007. ('Reply to the Letter of Duan Qirong et al., of Dongjing Village, Liuji Town Concerning the Situation of Julong Chemical Co., Ltd.'s Environmental Pollution', Funing County Environmental Protection Bureau).

24 Administrative Litigation refers to a situation in which a written legal complaint is made to the People's Court on the grounds that a citizen, legal person or other organization has not fulfilled its legal responsibilities. The Administrative Litigation Law of the People's Republic of China states that when citizens, legal persons or other organizations consider that administrative organs or staff have violated their legal rights, they can bring a suit to the People's Court under this law.

Even though the villagers did not win their case against the EPB, and the outcome of the lawsuit was not fully satisfying, it nonetheless led the EPB to decide to relocate the chemical plant, resulting in the villagers' only victory in their resistance to the factory since 2001. In January of 2008, when we went to Dongjing to investigate for the second time, some of Julong Chemical's workshops had already relocated to the County Chemical Industry Park, but some were still operating in Dongjing. In March 2009, on our third trip, Julong Chemical had been entirely relocated and on our fourth visit, in October 2011, rivers within the village had mostly returned to a clean state, and flocks of ducks could be seen feeding there (See Photograph 12).

After the EPB ordered Julong Chemical to move in 2006, the villagers felt it was unfair to let the plant off so lightly and just let it relocate, when they had endured years of pollution which had severely damaged their finances and their health. So Shu Qichang, Duan Qirong, Chu Yuhua, and Bian Xiaoren discussed suing Julong Chemical and demanding that the enterprise provide compensation to all the villagers for the damage it caused between 2001 and 2006. They went door to door to discuss this with the villagers, and those who agreed to participate in the compensation lawsuit stamped their personal seal on the documents and wrote down their personal identification numbers, so that if they won, they would get part of the compensation. In the end the group rallied 369 villagers.

When the plan of the villagers to join forces and file a lawsuit was discovered by the county and town governments, the county leaders 'invited them for a talk'.

> The people from the county government came to talk with us. They called it 'talking', but in fact they put us under house arrest [...] in the Wenming Hostel over there at the cotton mill. There were 11 police cars downstairs and 48 police officers. And the Liuji Township Party Secretary was talking with us upstairs, demanding that we give the materials to him. The materials they wanted were the documentation of the participants supporting our lawsuit [...] the government talked to us, and said they will compensate us 70,000 yuan. (Su Qichang Interview 1, August 2007).

The villagers refused the 70,000 yuan in compensation and refused to give the materials to the county government. On 18 September 2006, Duan Qirong, Shu Qichang, Chu Yuhua, and Bian Xiaoren filed their lawsuit with the Municipal Intermediate People's Court on behalf of 369 plaintiffs. In this civil complaint, the villagers' demands were that the court order the defendant to pay a total of 738,032 yuan in compensation for losses caused by industrial pollution.[25] The key to whether this lawsuit could succeed was whether the plaintiff villagers could provide evidence of damages. The villagers' agent, Zhang Dafu, thought very hard about how to do this, but in the end, the court determined that the villagers could not meet the preliminary burden of proof, and provide evidence of the existence of pollution damage, so the villagers lost the case. The villagers were dissatisfied by the decision of the Municipal Intermediate Court, and appealed to the Provincial High Court. The High Court also determined that the villagers could not meet the preliminary burden of proof and upheld the ruling of the lower court. The decision of this High Court was the final judgement.

Analysing these two lawsuits, we can see that the Dongjing villagers lost their case because they could not provide evidence that had legal effect. On the surface, the ruling of the court protected justice; the villagers could not provide evidence that had legal effect, and it was natural that they should lose the case. But we need to think more deeply about why the Dongjing villagers were unable to turn the 'social facts'[26] they saw and experienced into 'legal facts' which the court would recognize. Did they actually have the power to turn social facts into legal facts, and if they did not, then when the law did not take into consideration the villagers' ability to provide evidence, was it actually protecting justice or injustice?

With regard to the allocation of the burden of proof in compensation lawsuits caused by environmental pollution, the Supreme People's Court document, 'Some Provisions On Evidence in Civil Procedures' (2002), Clause Four, states: 'In compensation lawsuits for damages caused by environmental pollution, the violating party shall be responsible for producing evidence to prove the existence of exemptions of liability as provided in the law and that there is no causal relationship between the act and the harmful consequences.' However, applying this 'reversed burden of proof' in pollution disputes does not necessarily mean total exemption of the injured party from the burden of proof. The prerequisite for the injuring party providing

25 At the end of this civil complaint, the villagers explained how they arrived at the figure of 738,032 yuan: 'The plaintiffs request from the defendant 2,500 yuan each for the cancer victims' family members' psychological injury, hence 82,500 yuan. To compensate the 369 plaintiffs (189 households, and a total 678 individuals) 0.1 yuan per person per day for damage to their vegetables, totaling 538,200 yuan for 6 years; and to compensate the plaintiffs for the loss of 897 *mu* of paddy fields with an annual output of 500kg/*mu* and 0.1 yuan/0.5kg, a one-time payment to the villagers for the loss of 538,200 yuan over 6 years.

26 I distinguish 'social facts' and 'legal facts' because 'social facts' can only turn into 'legal facts' when they have been recognized by the law. Even if the 'social fact' is true, if it cannot be converted into a 'legal fact' through the path required by the law, it does not have any legal effect. The author's use of 'social facts' and 'legal facts' here is inspired by Chen Ajiang.

evidence is that the injured party meet a 'preliminary burden of proof' that they were harmed by pollution. But in real life, proving that pollution constitutes a hazard is beyond the capability of ordinary farmers. For example, in the 'Civil Judgement' of the Provincial High Court, the original statement concerning whether Dongjing had fulfilled the burden of proof was that they: 'were unable to provide evidence concerning a decrease in the quantity of crops, of vegetables being inedible, or of fish and shrimp dying from pollution, or any evidence to support the exact amount, cost, and scale of the loss incurred [...] they were unable to provide the medical records and proof of death of cancer victims, and provide expert conclusions regarding aetiology from the authorities.' The mechanisms through which pollution causes harm are incredibly complicated, and the scientific nature of the correlation between cancer, environmental pollution and the decrease in crop production, strange odours in vegetables, and fish and shrimp dying is still unclear. It is even less possible for ordinary farmers to provide proof of the mechanisms through which pollution generated such hazards. This hurdle of providing evidence therefore effectively shut a large number of victims outside the door of the law.

Aside from this, the lack of independence of the judiciary is another barrier that to some extent prevents justice from prevailing. Dongjing villagers were not capable of obtaining evidence themselves, so they filed an 'Assessment Application Letter' to the High Court in accordance with statutory procedure. In the application letter, the villagers requested the following: 'It is currently the season for the masses to harvest rice, and this is the best time for random sampling of rice. As such, we request that the High Court approve the assessment of material evidence regarding rice and water quality'. But the Provincial High Court made its judgement in a situation where there had been no evaluation of rice and water samples. The villagers and their lawyer were indignant at the High Court for making a ruling when the collection of evidence according to statutory procedures had not been completed. They believed the only reason why the High Court did not collect evidence was that it was afraid the test results would point to pollution issues with the water quality and rice. In fact, beyond the surmises of the villagers, there were also some practical judicial problems which caused the High Court to hold back on the ruling of this case. At the time, there were many pollution lawsuits under way in China, but because of the lack of scientific evidence of the mechanisms through which pollution causes disease, and the additional consideration that there was as yet no previous case in which the injured party had won when there was a lack of evidence, the High Court was reluctant to set a precedent.

Concluding Thoughts

From a case like Dongjing, we can see that the problem of the relationship between pollution and cancer is not a purely medical question; it is also a social question. The relationship exists under certain social conditions, and it cannot be entirely separated from the particular set of social institutions within which it is embedded, or from the specific socio-cultural context. Dongjing villagers engaged in resistance to environmental pollution for eight years. This was a 'protracted war' in which every initiative led to defeat and every defeat led to further resistance. It was this repeated and frustrating environmental resistance that brought the issue of pollution and cancer to centre stage. As it evolved, the case reflected the standpoint of the factory, the contradictions that local governments face in industrial development, the mechanisms for communication between the government and the people, the limited freedom and impact of the media, the dearth of medical research, and the powerlessness of the law when medical evidence is lacking. It is repeatedly frustrated environmental resistance like this that allows us to see which parts of the social system are in need of improvement.

In terms of the socio-cultural context, the Dongjing villagers' eight years of environmental resistance had many local characteristics. Currently, in many rural parts of China, villagers are constantly at risk of harm from industrial pollution because they did not persistently engage in resistance against it. The villagers' choice of tactics and the duration and impact of environmental resistance are strongly related to the characteristics of local village elites. The main organizer of Dongjing's resistance, Duan Qirong, was an elderly man over 60 years old, who had exceptional personal characteristics: he was just, tough, courageous, visionary, and rich in experience, and he loved to stand up against injustice on behalf of other people. He also had a strong legal awareness, and the ability to organize and mobilize others, without considering the personal cost to himself. During the eight years of environmental resistance, Duan Qirong and his descendants were repeatedly subject to intimidation and threats from the polluting factory, but he clung to one belief, that 'this matter affects future generations, and it won't do if there is not a strong resolve to drive the chemical plant away'.

Research into rural industrial pollution and the risks it presents must be understood against the backdrop of different stages of regional economic development. In reality, local governments have different levels of awareness and acceptance of environmental pollution at different stages of economic development. From 2000, counties and municipalities in the Subei region where Dongjing is located competed with each other to lower the cost of land and other investment and environmental standards in order to attract investment. There were media reports of a Subei town official stating that 'our advantage is that we are not afraid of pollution[27] 'in order to attract outside investment. Another government official in a local Investment Bureau in Subei explicitly remarked, 'We have enormous local environmental carrying capacity, and we are well within the limits prescribed by environmental indicators. The area is next to the sea, and contaminants can dissolve naturally, so there will be no environmental protection fees'.[28] In the context of this ideology and choice of development strategy, large numbers of seriously polluting enterprises were invited into Subei from more developed areas, with some clustered in industrial parks, and others scattered around the countryside. The phenomenon of every village having a polluting enterprise, as happened in Dongjing, was common, and pollution incidents occurred frequently. Through this type of development, the local economy grew rapidly.

During our fieldwork in October 2011, we discovered that as the economy grew, and the industrial structure gradually improved, Yancheng City began to be less dependent on intensely polluting industries, and there was a shift in the development ideology and understanding of the trade-offs between environmental protection and economic growth. First, the ideology behind attracting investment changed from 'inviting business and *attracting* investment' to 'inviting business and *choosing* investment', and the environmental threshold for entry was raised. In 2007, new chemical projects were required to make an investment of more than 50 million yuan; and in 2008, individual projects were required to make a onetime investment of at least 50 million yuan. From 2010, projects that required the expropriation of new land had to have an investment of at least 100 million yuan, excluding the cost of land, and this had to be made in a single instalment. Projects also had to pass environmental impact, safety, energy consumption and other assessments.[29] Second,

27 There are many related reports. See 'Frequent Environmental Clashes, 'GDP Complex' is the Culprit – A Subei Town Official Says in the Process of Attracting Investment: 'our advantage is that we are not afraid of pollution.' http://info.water.hc360.com/2005/09/12081758277.shtml and http://www.39.net/360_moved/fsjk/tpxw/132260.html. Accessed 21 January 2020.

28 Sun, 'China Pollution Migration Map: How Long Can Environmental Provisions Last – Sunan Pollution Heads to Subei'.

29 Information about environmental governance of Yancheng is from the document, 'Report on the Basic Information of the Municipal's Chemical Production Enterprises', 20 December 2010. Provided by Economic and Informatization Committee of Yancheng.

environmental pollution prevention and control efforts were stepped up. Under to the 'Work Plan for Deepening Regulation of Chemical Enterprises throughout the City', stricter regulation began in 2007 and continued until 2010. During these three years, 390 chemical enterprises were shut down, 59 relocated, and 44 transferred, representing 67.07% of all chemical enterprises when regulation began. The spatial concentration of chemical enterprises was also dramatically increased. By the end of June 2010, there were a total of 362 chemical production enterprises in the municipality, of which 269 were located within industrial parks and controlled areas, and 93 were outside industrial parks, for an aggregation rate of 75%. It is noteworthy that an employee of the County Chemical Production Management Office told us that the performance of Julong Chemical, which relocated to the industrial park, was superior to other enterprises; that the company had adopted advanced production technology and did not make any illegal emissions. From this we can see that the enterprise improved as the result of the villagers' resistance and regulation by the government.

From the case of Dongjing in Subei, we can also draw the following insights, which have significance for development in other regions. First, compared with other social problems, environmental health risks have special properties. Their uncertain nature means that there is the possibility of a risk to health or even life, and so it is logical to take active preventive measures and not wait passively for incidents to occur before dealing with them. But at the time of this research the official attitude tended to be to wait, and act only when problems occurred. In October 2011, the research team interviewed an employee in the County Centre for Disease Control about this. He told us that when it came to issues of pollution and disease, they would typically only go to investigate when a problem came up. This was clearly not a preventative approach. Second, in the environmental governance of China, chemical enterprises have become like 'rats crossing the street'. Everyone yells for them to be killed, but realistically, the chemical industry cannot just disappear; our everyday lives can no longer be disentangled from the use of all kinds of chemical products. And so our attitude towards chemical enterprises should leave them enough room for survival, and focus on regulating pollution through improved management and supervision, instead of making them, 'run every time they are shot at'. This just results in their relocating from developed areas to poor places with lower environmental thresholds for entry and taking their pollution and associated health risks with them.

Works Cited

Chen, Ajiang. 'Analysis of Stakeholders in Incidents of Water Pollution'. *Zhejiang Academic Journal*, 4 (2008): 169-175.

—. *Secondary Anxiety: A Social Interpretation of Water Pollution in Taihu*. Beijing: China Social Sciences Press, 2009.

Dong, Zhiwei, and Qiao Youlin. 'Report on China's Cancer Control Strategy'. *China Tumor*, 5 (2002): 250-260.

Fei, Xiaotong. 'Small Town, Subei First Exploration'. In *Small Town Four Notes*, 76-77. Beijing: Xinhua News Agency, 1985.

He, Xuefeng. *New Rural China – Research Notes on Rural Society During the Transition Period*. Nanking: Guangxi Normal University Press, 2003.

Honig, Emily. *Creating Chinese Ethnicity: Subei People in Shanghai, 1850-1980*. New Haven, CT: Yale University Press, 1992.

Jiangsu County Statistics webpage. *Jiangsu Statistical Yearbook – 2010*, electronic version, http://www.jssb.gov.cn/jstj/tjsj/tjnj/. Accessed 10 January 2013.

Jiangsu local blog network: http://www.jsdfz.com/Article.asp?articleid=388. Accessed 6 May 2008.

Li, Jiang, Ou Zhengtao, and Cai Yugao. 'GDP Complex is the Culprit – A Subei Town Official Says in the Process of Attracting Investment: "Our Advantage is that We Are Not Afraid of Pollution"'. *Banyuetan* 17 (2005): 36. http://info.water.hc360.com/2005/09/12081758277.shtml and http://www.39.net/360_moved/fsjk/tpxw/132260.html. Accessed 21 January 2020.

Olson, Mancur. *The Logic of Collective Action: Public Goods and the Theory of Groups* (Revised ed.). Cambridge, MA: Harvard University Press, 1971 (1965).

Pang, Jiaoming. 'Jiangsu Province Yancheng City Cancer Village: Pollution Causes Hundreds to Die of Cancer in the Previous 3 Years', 2 November 2005. http://news.sohu.com/20051102/n227363025.shtml. Accessed 21 January 2020.

Sun, Zhan. 'China Pollution Migration Map: How Long Can Environmental Provisions Last – Sunan Pollution Heads to Subei'. *China Newsweek*, 23 January 2006.

4 Environmental Change and Health Risks[1]

Chen Ajiang

Abstract

Cause of death data show that Lianshui County has a high incidence of oesophagus cancer. This chapter considers health risks in the county from the perspective of environmental history. After the Yellow River became blocked and took over the course of the Huai River, it caused Lianshui to suffer an 800-year period of floods, waterlogging and attendant disasters. This brought prolonged poverty and water-related health risks. The later conversion of dry fields to rice paddy brought further changes in the water environment of Lianshui's villages, connecting drinking water sources to the agricultural irrigation system, and causing new health risks. However, as the economic circumstances of the village improved, so did the drinking water supply, and water-related health risks then declined.

Keywords: environmental history, diet, rice paddies, ponds, environment and health, village doctors

During fieldwork in Lianshui, Jiangsu, I learnt that there was a high incidence of oesophageal cancer in some rural areas of the county. Later, research at the Health Bureau and a review of the literature confirmed this. In 2013, oesophageal cancer was the primary cause of death from cancer in Lianshui. The mortality rate was 49.94/100000, accounting for 34.19% of all cancer deaths,[2] and this was approximately three times the national rural mortality rate for oesophageal cancer for that period.

1 This article was published in *Xuehai* 2017 (issue 1): 73-83.

2 Sun, 'Analysis of 2013 Lianshui County Resident Malignant Tumor Deaths'. The oesophageal cancer mortality rate of Lianshui County provided by Sun Weixin is around 50/100,000, which is very different from the data provided by Li Qianwen et al. Li claims 'the morbidity rate

Chen Ajiang, Cheng Pengli, and Luo Yajuan, *Chinese "Cancer Villages": Rural Development, Environmental Change and Public Health*. Taylor & Francis Group 2020
DOI: 10.5117/9789089647221_CH04

The latest authoritative medical textbooks attribute the risk factors for cancer to two main sources:[3] environmental factors and hereditary factors. Environmental factors mainly fall into three categories: chemical carcinogens, and physical and biological factors. Hereditary factors include cancers that are passed on genetically within families but also hereditary susceptibility to cancer.[4] Purely hereditary cancers account for only a small percentage of the total and due to the interaction between genes and the environment, most human cancers are still caused by the pathogenic effect of environmental factors.[5] The precise causes of oesophageal cancer are as yet undetermined, but a substantial volume of statistical, experimental and intervention medical research shows that the main risk factors for the disease include environmental factors, nutrition, lifestyle, and genetics.

This chapter uses an environmental history perspective to analyse the effects of changes in the water environment on the production activities and lives of villagers in Lianshui County and examine how environmental change contributes to different types of health risks.

Literature Review

Oesophageal cancer is a common cancer, but its incidence varies considerably across countries. China has a high incidence of the disease. The Survey of Cancer Deaths in China, 1973-1973, shows that the total mortality rate for oesophageal cancer at that time was 19.09/100,000, ranking second among all cancer deaths; the 1990-1992 survey found that the oesophageal cancer mortality rate was 15.02/100,000, ranking fourth.[6] Oesophageal cancer in China has obvious regional characteristics. It is concentrated in parts of the

of oesophageal cancer in the Huai'an region from 2009 to 2010 is 72.48/100000, ranking first among malignant tumours in Huai'an. The morbidity rate in Lianshui County is the highest at 120.67/100000'. (See Li and Yuan et al., 'Analysis of Huai'an Region Esophageal Cancer Epidemic Characteristics and Treatment Situation', pp. 142-145). According to the various data and comparisons we acquired, the mortality rate of Lianshui County oesophageal cancer being 50/100000 should be closer to the truth. Therefore, data relating to Lianshui County cancer in this article is mainly based on statistics from Sun Weixin.

3 Also known as risk factors, i.e. factors that increase the possibility of the onset of certain diseases, and are thus called causative factors for that disease, or 'the cause'. In epidemiological studies, undetermined possible causative factors are often referred to as risk factors. Wang. ed., *Epidemiology*, p. 32.

4 Hao and Wei, eds., *Oncology*, pp. 14-16.

5 Ibid, pp. 15-16.

6 Wang, ed., *Epidemiology*, p. 261.

country that include the border area between Henan and Hebei, Northern Sichuan, the Dabie Mountains, and Northern Jiangsu. Existing research suggests that the main risk factors for oesophageal cancer include:[7]

1 Nitrosamines: Nitrosamine compounds are highly carcinogenic. The levels of nitrosamines and their precursors in the drinking water and food of regions with a high incidence of oesophageal cancer are significantly higher than in low-incidence areas. Nitrosamines are broadly carcinogenic and have been used to induce malignant tumours in all the tissues and organs of fish, amphibians, rats, rabbits, dogs, apes and other species.
2 Mould: mould contamination is significantly higher in the food of populations with a high incidence of oesophageal cancer, especially in pickled foodstuffs.
3 Unhealthy lifestyle: long-term smoking and alcohol consumption are related to the onset of oesophageal cancer.
4 Dietary habits: preferring foods that are too hot, eating too fast, and consuming food that is very hard or coarse may be related to oesophageal cancer.
5 Nutrients and trace elements: lack of animal proteins, fresh vegetables and fruits (lack of essential vitamins B2, C, and A, and essential fatty acids) and micronutrients such as molybdenum, zinc, iron, copper, manganese may be related to oesophageal cancer.
6 Genetic factors: oesophageal cancer displays relatively obvious familial clustering. Although this is mainly due to the effects of a shared living environment, genetic factors cannot be completely ruled out.

The geographical concentration of oesophageal cancer in China was noted as early as the 1950s in places like Lin County, Henan and Yanting County in Sichuan, and this prompted the government to set up the death registration system, conduct scientific research, and carry out disease prevention and control. Oesophageal cancer research and prevention in Lin County, Henan is very representative of this work, and the problem in the county was a matter of concern to Premier Zhou Enlai and other state leaders. Systematic cancer death registration and scientific research initiated in these early times have provided quite comprehensive statistics on oesophageal cancer in Lin County, and the research was fairly in-depth. Research into the aetiology of oesophageal cancer in the area suggested the causes included: foods such

7 See Gao, ed., *Experts Talk on Countering Esophageal Cancer*, pp. 199-200. Also see Xu, ed., *Fertilizer, Waste Water, and Esophageal Cancer*, pp. 16-118.

as pickled cabbage and harmful mould on food; lack of micronutrients in the natural environment; unhealthy lifestyles; and the 'nitrogen cycle hypothesis' which centred on water pollution.[8]

The 'nitrogen cycle hypothesis', which was put forward by Xu Zhixiang, has received partial verification from medical statistics, experiments, and evidence from water improvement projects. The hypothesis focuses on the development of nitrosamines, explaining how their precursor substances evolve in the environment and enter the human body. The main points of the 'nitrogen cycle hypothesis' are as follows:[9]

1. Nitrogen-nitroso chemical compounds which include nitrosamines and nitrosamides are broad spectrum carcinogens, which can induce tumours in all organs of a range of animal species. Nitrosamines with a symmetrical chemical structure cause oesophageal cancer, while those with an unsymmetrical structure cause liver cancer; nitrosamides cause stomach, oesophagus, and other cancers.
2. Naturally occurring amines and amides and nitrates and nitrites in waste water enter the human body through drinking water, where they combine to form *N*-nitroso compounds. Amines and amides mainly originate from rural compost and waste water from paper mills etc.
3. The hypothesis also attempts to integrate the risks factors of mouldy foodstuffs, and the absence of nutrients and micronutrients into the 'nitrogen cycle hypothesis'[10].

The progression from the presence of various chemical substances in the environment to disease risks in the 'nitrogen cycle hypothesis' is shown in the following diagram.

The 'nitrogen cycle hypothesis' has been partially verified through experiments on animals in 1989, Xu Zhixiang and Chen Fenglan and used nitrosified agricultural fertilizer water to induce oesophageal cancer in two out of 63 chickens. In 1999, a replication of the experiment in which 100 chickens were fed nitrosified agricultural fertilizer water induced oesophageal cancer in 16 of them, stomach cancer in five, and liver cancer in three. None of the chickens in the control group developed cancer. Previously, although attempts had been made since the 1930s, researchers in China and abroad had failed to induce oesophageal cancer using substances including

8 Xu, ed., *Fertilizer, Waste Water, and Esophageal Cancer*; Xu, Tan, Han, and Chen, 'Esophageal Cancer, Stomach Cancer, Liver Cancer Nitrogen Cycle Hypothesis and Testing', pp. 5-7.

9 Ibid.

10 Ibid.

Figure 4.1 Oesophageal cancer nitrosamine risk diagram

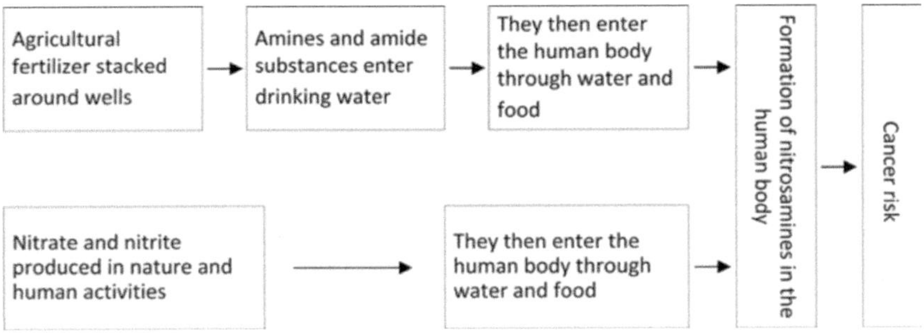

acetamido-fluorene, methylcholanthrene, vomit from oesophageal cancer patients, mouldy cheese, sodium nitrite, and nitrosamines.[11]

Research also found a noticeable decrease in the incidence and mortality rate of oesophageal cancer in some populations in Linzhou City after the source of drinking water was improved (that is, they were provided with running tap water). Han Jianying et al. investigated 344 villages in 10 townships in Linzhou between 1996 and 1998 and found that the oesophageal cancer and mortality rates in the populations which had undergone standardized improvement of the water supply were noticeably lower than those in communities with nonstandard improvement or no improvement. The incidence decreased by 42.10% and 51.59% respectively, and the mortality rate decreased by 46.77% and 49.08%.[12] Similarly, Cheng Lanping et al. discovered that the oesophageal cancer morbidity rate was different for populations with different types of drinking water supply, decreasing significantly in the following order: well water, mountain spring water, artificial well water, motor pumped well water, river water, and a standardized water supply (reservoir water).[13]

Most of the published research on oesophageal cancer in Lianshui provides statistical analysis of cancer incidence and mortality rates;[14] so far,

11 Xu, Tan, Han and Chen, 'Esophageal Cancer, Stomach Cancer, Liver Cancer Nitrogen Cycle Hypothesis and Testing'; Xu, Tan, Chen et al., 'Farm Compost polluted Water May Induce Pharyngo-Esophageal, Gastric and Liver Carcinoma in Chicken', pp. 344-347.

12 Han, Lian and Xu et al., 'Influence of Standardized Deep Water on Morbidity and Mortality of Esophageal Cancer in Linzhou', pp. 200-202.

13 Cheng, Lian and Liu et al., 'Influence of Water Quality Improvement on Incidence and Mortality with Esophageal Cancer in Linzhou, Henan', pp. 998-999.

14 This may be related to the death registration system which was only initiated in 2009. Before this, there was not a complete death statistical data. Even though the researched Lianshui County was a high incidence county in 1973-1975, it was not the most representative, thus Lianshui County was not in the sample survey of 1990-1992. See Sun, 'Characteristics of 2001

there are no very useful aetiological studies. Sociological and anthropological research on cancer in China has focused on the relationship between people's awareness, actions, social structure and cultural heritage and their health. The research of Anna Lora-Wainwright in a village in Langzhong, Sichuan focused on analysing villagers' perception of cancer.[15] But because the medical facts about cancer were unclear, there was no way to make a judgment about the incidence of cancer and this affected Lora-Wainwright's ability to make a compelling analysis of this controversial 'cancer village'. In fact, cancer research from the 1970s had already demonstrated that Northern Sichuan is one of the regions in China that has a high incidence of oesophageal cancer; for example, the disease is prevalent in Yanting County, which is near Langzhong.

A similar lack of clarity is evident in international research. In Martha Balshem's study of Tannerstown in Philadelphia, USA, there were clear differences in the factors to which different actors attributed cancer. The government assumed that cancer originated from lifestyle choices, and, on the basis of this, the 'CAN-DO' cancer education program attempted to change residents' behaviour in order to reduce their risk. But residents attributed the cancer risk more to environmental pollution, which the government and scientific authorities neglected. Participants in 25 interviews and eight focus groups chose environmental factors (environmental pollution, food additives, and so on) that were beyond personal control twice as frequently as they chose preventable or lifestyle factors which cause cancer.[16] It is very natural that the underlying uncertainty regarding the causality of cancer, combined with differences in perceptions due to social background and access to scientific information, leads to differences in perceptions which sometimes result in disputes.

Starting out from the Chinese context, my research took a different direction from that of Martha Balshem. When conducting research on the 'cancer villages' of Shenqiu County, Henan, I found that when pollution of the Huai River Basin became a social problem that attracted a lot of attention, the health risk factors which were internal to the village were completely overlooked. Both my own research and that of my collaborators on 'cancer villages' in other locations shows that aside from scientific knowledge, the

High Incidence of Lianshui County Resident Malignant Tumor'; Sun, 'Analysis of 2013 Lianshui County Residents Malignant Tumor Deaths', pp. 34-35.

15 Lora-Wainwright, 'Anthropological Study of Cancer Villages: Perception of Villagers on Accountability and Coping Strategies', pp. 238-262.

16 Balshem, *Cancer in the Community*, p. 71.

stage of economic development and the socio-cultural background of the residents are also important in shaping their awareness and responses to the problem of the relationship between pollution and cancer.[17]

In fact, both the health problems of Lin County (Linzhou) and the problem of oesophageal cancer issue in Lianshui County discussed in this chapter share a very important dimension, and that is environmental change. Environmental history is therefore an important perspective in understanding these cases. In recent times, the massive scale and rapid pace of environmental change have transformed the world and become a vibrant topic of research. Mark Elvin's book, *The Retreat of the Elephants*, provides readers with a vast and evocative overview of thousands of years of environmental change in China, and explains that the retreat of elephants from Northern China to the South did not happen primarily because of climate change, but rather because of the human destruction of forests and other habitats suitable for elephants.[18] Robert Marks examined in detail the interactive process between social development and environmental change in the Lingnan region from the early Ming Dynasty to the late Qing Dynasty.[19] Both Elvin and Marks focus only on change in the environment and ecosystems, but their systematic investigations are also significant and enlightening for research on environment and health.

Research Methods

This chapter explores the high incidence of oesophageal cancer in rural Lianshui from the perspective of environmental history. Problems such as the high incidence of oesophageal cancer are usually considered to be topics for medical research, but the social sciences can also make a contribution. Studies of the aetiology of oesophageal cancer are not only investigations into chemical substances and the biological world. These are also interwoven with production activities, lifestyles, and even social structure and cultural heritage. This is clear in the history of epidemiology. For example, in 1774, the British surgeon Percival Pott discovered that the pathogenesis of scrotal cancer was closely related to the occupation of the patients, who were chimney sweeps. Another example is research that found a close relationship between smoking and respiratory system cancers. In research on oesophageal

17 Chen, 'Inside and Outside 'Cancer Village'.
18 Elvin, *The Retreat of the Elephants – China's Environmental History.*
19 Ma, *Tigers, Rice, Silk, ad Silt: Environment and Economy in Late Imperial South China.*

Figure 4.2 The relationship between oesophageal cancer and risk factors

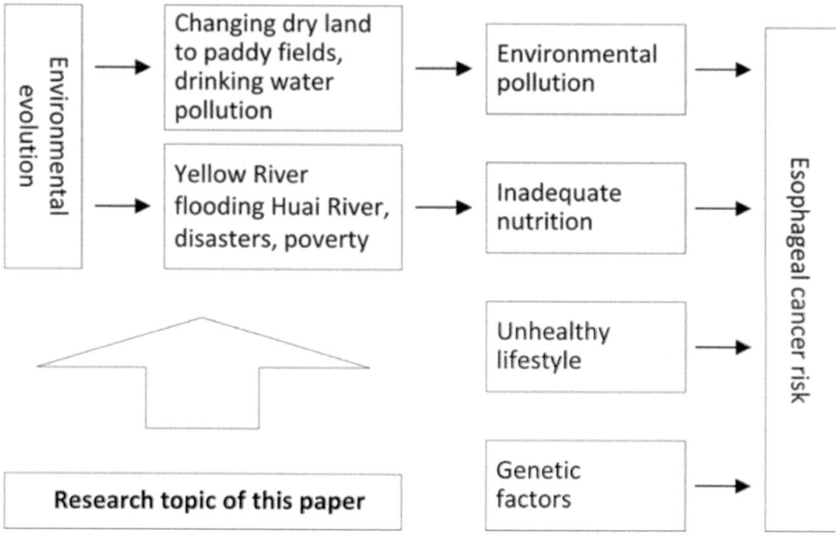

cancer in Lin County, the relationship between nitrosamines and cancer may be purely a matter for bio-chemistry and biomedical science, but the production and use of agricultural fertilizer, and the process through which amides contained in fertilizer enter drinking water are questions for social science. Medical and social science research are thus complementary, and this is clear in the integration of sociology, anthropology and psychology into the study of the causes and treatment of disease. Since the 1970s when G.L. Engel proposed the biopsychosocial model of medical science, comprehensive models of this type have been widely accepted[20].

In positioning themselves in interdisciplinary environmental health research, sociology and other social sciences need to identify how their particular perspective can inform the understanding of the issue and find a standpoint from which to conduct research. Clearly, this is different from ordinary, 'pure' social science research; researchers must identify their own research problems within intricate, interdisciplinary 'problem clusters'. Achieving a comprehensive understanding of these 'problem clusters' and 'dissecting' the social science problems that lie within them therefore becomes the key task.

It was with these methodological considerations in mind that I began conducting research on the relationship between changes in the water

20 See Zhang ed., *History of Medicine*, pp. 220-222.

environment and the high incidence of cancer in rural Lianshui. The risk factors for oesophageal cancer in the county included environmental pollution, nutritional issues, lifestyle, and genetics. In terms of environmental pollution, the conversion of dry land to paddy fields starting in the 1970s had a major effect on drinking water sources, and imbalanced dietary nutrition was related to the long periods of poverty in the region, which was in turn closely linked to the environmental changes that resulted from the Yellow River flooding the Huai. There is no literature suggesting that the population of Lianshui is unusual in terms of lifestyle or genetics and so these factors are not discussed here. Instead, as shown in Figure 4.2, I focus on environmental changes, (the flooding of the Huai by the Yellow River), and changes in agricultural practices (the shift to rice cultivation).

The analysis draws on fieldwork and a review of the relevant literature. Fieldwork in Lianshui in October 2014 included:

1 interviews with clinic doctors in three administrative villages: Bei Village, Hua Village, and Xin Village;
2 detailed observation and consultations regarding the layout and water systems of Bei Village, and of Ni Village in Zhuma Township (which historically had a high incidence of cancer);
3 interviews with the vice director of Qianji Township's Disease Prevention and Healthcare Office to understand cancer morbidity and his views on the high incidence of oesophageal cancer, as well as to collect data on cancer for the whole of Qianji Township;
4 interviews with individuals from the Lianshui County Health Bureau's Disease Prevention and Healthcare Department and the Lianshui County Centre of Disease Control (CDC) to understand their perspectives on health data, diseases and their possible causes, and changes in the local environment, especially in the water system.

Cancer statistics for Bei Village were provided by Mr. Wang. Data for 2011-2013 are those reported to the death registration system by the village clinic; data for the period before that were obtained by interviewing villagers about their recollections. Data on the evolution of the water system, natural disasters, and the economy came mainly from the 'Lianshui County Gazetteer' and the 'Lianshui County Water Conservancy Gazetteer'.

A review of the literature on Lin County (Linzhou) and other counties and cities with a high incidence of oesophageal and other cancers and other counties gave me a fairly comprehensive understanding of the complexity of oesophageal cancer, and its relationship to environmental factors, especially the water environment. This literature included medical research and more general background literature on the geographical environment and socio-economic development, for example the 'Lin County Gazetteer' and 'The Red Flag Canal Chronicles'.

The High Incidence of Oesophageal Cancer

From 1975 to 1978, the Ministry of Health Cancer Prevention Research Office organized a three-year (1973-1975) mortality census in 29 provinces, municipalities, and autonomous regions in China. It covered more than 850 million people, but due to transport problems, 35 counties in Tibet and Sichuan were not investigated.[21] This research shows that of the 900 million people in China at the time, around 700,000 died of cancer each year, and the three main types were stomach cancer (160,000), oesophageal cancer (157,000), and liver cancer (100,000). Jiangsu Province had an annual cancer mortality rate of over 70,000, the highest of any province in China.[22] According to statistics from the World Health Organization cited in that investigation report, at that time, China had the highest oesophageal cancer mortality rate at 23.40/100,000. Puerto Rico ranked second, but was far behind, at 9.31/100,000; Oesophageal cancer was also the primary contributor to cancer mortality, at 23.53%.[23] Henan Province had the highest oesophageal cancer mortality rate at 32.22/100000, Jiangsu was in second place with 29.22/100000; oesophageal cancer accounted for 40.55% of cancer deaths in Henan Province and 26.77% in Jiangsu Province. Areas with higher oesophageal cancer mortality rates had particular geographical characteristics. In Jiangsu, the Subei (northern Jiangsu) region, which includes Huai'an County,[24] was an oesophageal cancer hotspot, and Lianshui County was also on the list of high mortality counties.[25]

21 Ministry of Health, Office of Cancer Prevention and Treatment, ed., *China Malignant Tumor Mortality Investigation*, pp. 1-3.

22 Ibid., p. 11.

23 Ibid., p. 73.

24 In the 1970s when the investigation was conducted, Huai'an County was a county under the jurisdiction of Huaiyin District in Jiangsu Province. In 1983 Huaiyin District was revoked and Huaiyin City was established, under the jurisdiction of Jiangsu Province, and Huai'an County belonged to Huaiyin City. In 1988 Huai'an County was revoked and county level Huai'an City was established. In 2001, Huaiyin City was renamed Huai'an City, the initial county level Huai'an City was revoked to establish district, renamed Chuzhou District.

25 Ministry of Health, Office of Cancer Prevention and Treatment, ed., *China Malignant Tumor Mortality Investigation*, p. 96, and p. 98.

At the end of 1992, the National Cancer Prevention and Treatment Research Office organized a random survey on the cause of death for the three years from 1992-1993; the sample covered 10% of the total population. The results for 22 provinces found the cancer mortality rate to be 106.7/100,000. The most prevalent cancers were, in descending order: stomach (25.2/100,000, or 23.2% of all cancer deaths), liver (20.4/100,000, or 18.8%), lung (17.5/100,000, or 16.2%), and oesophageal (17.4/100 000, or 16.1%).[26] In eleven provinces the oesophageal cancer mortality rate was higher than the national average, including: Shanxi (42.5/100 000), Henan (34.6/100 000) and Jiangsu (30.0/100 000). Jiangsu still placed third, so the situation was not much different from the investigation twenty years prior. There were eight counties (including cities and districts) in Jiangsu where the sampled cases exceeded the national standard by more than 100%. These included Huai'an City, (91.8/100 000), and Dafeng County (46.1/100 000). Lianshui County was not a sample county, so there is no data.

Like other counties, in 2009 Lianshui began the registration of deaths from chronic disease. After a few years of experimentation and organization, the system began normal functioning in 2011. According to statistics provided by Sun Weixin, in 2013 oesophageal cancer placed first among cancer deaths in Lianshui. The mortality rate was 49.94/100 000, or 34.19% of all cancer deaths,[27] and nearly three times the national rural oesophageal cancer mortality rate for the period.[28]

We invited the village doctor Mr. Wang to help us conduct research in Bei Village by obtaining statistics from 1990 to 2013. During this 24-year period, he found that a total of 91 people contracted cancer, of whom 53 had oesophageal cancer, or a total of 58.24%. Forty were male, and thirteen were female, a ratio of one to three. Given that Bei Village had a population of 2500, the cancer morbidity rate for those 24 years was 151.67/100 000, and for oesophageal cancer it was 88.33/100 000. Both figures are high. We also analysed preliminary statistics for the two villages next to Bei Village in order to check the disease registration data for Qianji Township where

26 National Tumour Prevention Office, *China Malignant Tumor Mortality Investigation (1990-1992)*, p. 26.

27 Sun, 'Analysis of 2013 Lianshui County Residents Malignant Tumor Deaths', pp. 34-35.

28 The data that Lianshui County oesophageal cancer's mortality rate is 50/100,000 is provided by Sun Weixin, and has large differences from the statistics provided by Li Qianwen et al. Li claims that 'the morbidity rate of oesophageal cancer in the Huai'an region from 2009 to 2010 is 72.48/100,000, first place among Huai'an City's malignant tumours. Lianshui County's morbidity rate ranks the highest, at 120.67/100 000'. (See Li, Yuan, Du et al., 'Analysis Research of Huai'an Region Esophageal Cancer Epidemic Characteristics and Treatment Situation', pp. 142-145). According to comparison with the many types of statistics which we obtained through research, the mortality rate of 50/100 000 for oesophageal cancer in Lianshui County should be closer to the truth.

Graph 4.1 Number of people with oesophageal cancer in Bei Village

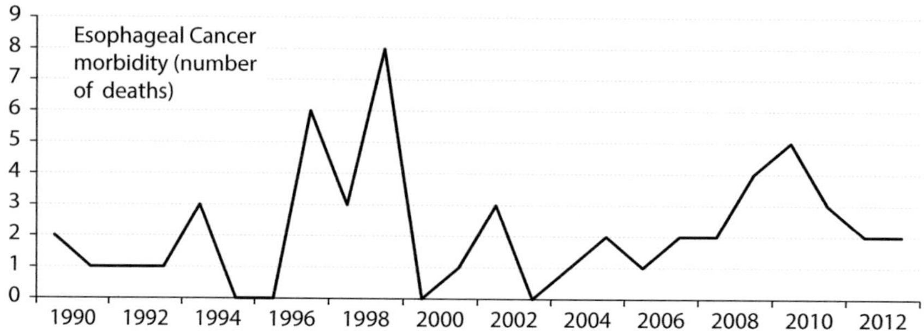

Bei Village is located. The data for these villages was very similar to that for Bei Village.

From these statistics we can see that over the twenty years prior to our research, the oesophageal cancer morbidity rate in Bei Village did not change much. Even though there were annual differences, if we take five and ten-year periods as the unit of analysis, the overall change is small. In the next section, we consider the ways in which environmental change may have contributed to these high rates of oesophageal cancer.

The Yellow River Flooding the Huai, Natural Disasters and Poverty

Lianshui County is situated to the north of Huai'an City and falls under its jurisdiction. The county lies at the point where the old Yellow River and Huai River enter the ocean. In 2013 Lianshui County's registered population was 1.1 million and the county covered 1,679 square kilometres.[29]

The Huai River originates in the Tongbai Mountain in Henan Province, and flows through Henan and Anhui before entering Jiangsu. Before the Yellow River started flooding it (prior to 1194), the Huai flowed northeast towards Huaiyin through the Gui Mountains. There were only a few scattered small lakes nearby, and the modern Lake Hongze did not exist. In Huaiyin (now Huai'an), the tributaries Sishui and Yishuijo enter the Huai, and the Shushui joins it at Lianshui, The Huai flows towards the east, roughly following the current southern border of Lianshui County in the Shihu region and then exits at Yuntiguan and enters the sea.

29 http://www.lianshui.gov.cn/web/center/2281/9069/9070/9070.shtml. Accessed 5 January 2015.

The section of the Huai that flows through Jiangsu is long and deep and before it was flooded by the Yellow River it entered the sea smoothly, so during that time there are not many records of flood disasters in Lianshui, which was close to the sea.[30] But after 1194, and especially after 1495 when the Yellow River overflowed and flooded the Huai, the countless floods, waterlogging, and droughts and other secondary disasters that resulted changed the environment of people living along the Huai, and along with it their lives and their fate.

1194 was a turning point. In that year, the Yellow River overflowed its banks at Wuyang in Henan. After this, one branch entered the sea through Beiqing River (now the Daqing River), and the other entered the Huai through the Nanqing River (now called the Sishui). For three hundred years between 1194 and 1945, the Yellow River ran through the tributaries of the Huai, including the Wo, the Ying, the Bian and the Si. But prior to 1495, although it had quite a big impact on the upper and mid-stream sections of the Huai, this did not yet constitute a very severe risk to the downstream region of Lianshui.[31]

In 1495, due to the successful damming of the Yellow River at Huanglinggang, the Yellow River's path into the sea through the Beiqing River was blocked, and Yellow River overflowed into the Huai through the Sishui tributary at the current intersection of Xuzhou, Suqian and Huai'an in Jiangsu. The channels of the Huai River below Qingko were flooded. The Yellow River carried a large amount of mud and silt and, over time this raised the riverbed to the point where it blocked the exit to the sea. Water flooded Subei, and Lianshui County bore the brunt of the deluge. During this period, the county suffered frequent disastrous floods.[32]

In 1855, the Yellow River overflowed its banks again. It changed its path at Tongwaxiang in Henan Province, flowed northeast, and entered the sea at Lijin in Shandong. This ended the direct threat from floods in Lianshui. But the Yellow River's flooding of the Huai and multiple changes of course had raised the river bed and blocked the Huai's exit to the sea at Lianshui. Water backed up, causing floods. Because of these disasters, the people had no way of making a living, the population decreased, and in the Ming Dynasty it was twice suggested that the county should be abandoned.[33]

After Liberation, Lianshui suffered from another problem, salination. In 1951-1995 the average annual rainfall in Lianshui County was 981.3

30 Lianshui County Water Conservancy Gazetteer Editorial Committee, *Lianshui County Water Conservancy Gazetteer*, p. 44-45

31 Ibid., p. 45.

32 Ibid., p. 45.

33 Ibid., p. 4.

millimetres, and from 1952-1995, the average evaporation rate was 1415.7 millimetres. This situation, in which the evaporation rate exceeded precipitation rate, provided the basic climatic conditions for soil salination. As accumulated surface water continuously evaporates, ground water rises, and salt builds up on the surface, causing severe salination. According to statistics from the 1950s, the total arable land in Lianshui was 226.93 *mu* (including a part of the today's Guannan), of which salinized wasteland accounted for 62 *mu*, or 36.6% of the total area.[34] This affected the productivity of the soil. Lianshui's food crops were primarily wheat, corn, potato, and soybeans, and aside from potatoes, the yields were less than 50 kilograms.[35]

According to historical records, in the hundreds of years before the Yellow River flooded the Huai, natural disasters were quite rare in Lianshui County, occurring approximately once every 50 years. In the three hundred years between the Yellow River flooding the Huai in 1195 and the blocking of the Yellow River at Huanglinggang (1195-1495), disasters increased, especially floods, which occurred about once every ten years. In the half century between the blocking of the Yellow River at Huanglinggang and flooding at Caowan (1495-1552), floods occurred on average once every four years, and droughts also greatly increased, averaging one every seven years. In the three hundred years between the flooding at Caowan and the Yellow River migrating north (1553-1855), disasters were at their worst, occurring almost once a year, with floods once every two years, and waterlogging once every four years. The frequency of droughts also increased, occurring once every five years. In the hundred years between the Yellow River migrating north and the founding of the People's Republic of China (PRC) (1856-1949), the frequency of floods decreased, but waterlogging and droughts did not, and the overall frequency of disasters was still very high, reaching three every four years (see Table 4.1 below.) In the half-century after the founding of the PRC the construction of irrigation works decreased the number of floods noticeably, but the frequency of waterlogging was still relatively high, although this may be related to the level of detail in the records. Regardless, the severity of corresponding disasters was controlled.

Historically, famines caused by floods, waterlogging, and drought were common everywhere in China. For example, an account of the great floods of Year 32 of Guangxu (1906) reads, 'From the mid-autumn festival to the moon festival, there was excessive rain for 70 days [...] the second harvest of wheat all rotted [...] people have to eat wheat saplings, weeds, elm bark, poppies, Guanyin mud and so on. In the Xu, Hai and Huai areas there are about

34 Ibid., p. 203.

35 Lianshui County Gazetteer Editorial Committee, *Lianshui County Gazetteer*, pp. 244-245.

Table 4.1 Floods, waterlogging and drought in Lianshui County

Time period	*Occurrence*	*Flood*	*Waterlogging*	*Drought*	*Tides (潮)*	***Total***
The 1006 years (188-1194) before the Yellow River flooded the Huai	Occurrences	12	3	7	2	**24**
	Hundred-year frequency	1.19	0.30	0.70	0.20	**2.39**
The 299 years (1195-1494) between the Yellow River flooding the Huai and the Yellow River being blocked at Huanglinggang	Occurrences	26	7	9	1	**43**
	Hundred-year frequency	8.70	2.34	3.01	0.33	**14.38**
The 57 years (1495-1552) between the Yellow River being blocked at Huanglinggang and the Caowan flooding	Occurrences	9	4	8	2	**23**
	Hundred-year frequency	15.79	7.01	14.03	3.51	**40.34**
The 302 years (1553-1855) between the Caowan flooding to the Yellow River migrating north	Occurrences	147	78	55	15	**295**
	Hundred year frequency	48.67	25.82	18.21	5.00	**97.70**
The 93 years (1856-1949) between the Yellow River migrating north and the founding of the PRC	Occurrences	21	25	19	5	**70**
	Hundred-year frequency	22.58	26.88	20.43	5.38	**75.27**
The 46 years (1949-1995) between founding of the PRC and 1995	Occurrences amount	3	29	9	–	**41**
	Hundred-year frequency	6.52	63.04	19.56	–	**89.12**
Total time period (1807 years)	Occurrences	218	146	107	25	**496**
	Hundred-year frequency	12.06	8.08	5.92	1.38	**27.44**

Note: the statistics in this table are calculated based on the Local Gazetteer Editorial Committee, Lianshui County Water Conservancy Gazetteer, Jilin Literature and History Publishing House, 2003, p. 65, 'Table 1.4.1 Lianshui County historical statistics on floods and droughts'. Because the records are detailed but brief, the data only has relative significance.

2,400,000 or 2,500,000 calamity victims [...] about 30,000 died of hunger or disease'.[36] There is also a record showing that in the most extreme situations, in both the Ming and Qing Dynasties, people turned to cannibalism.[37]

Statistics show that in the 30 years before reform and opening up, poverty was widespread in rural areas of Lianshui County and people suffered from malnutrition. There was no secure supply of staple foods. For example, in 1950, total grain production in Lianshui County was 77,442,000 kilograms, and the population was 462,000, so the average supply per person was only 167.6 kilograms. Of this low average grain production, coarse cereals accounted for a high proportion: 39.77% were three types of wheat, 19.87% was corn, 21.73% was potatoes, 10.63% was soybeans, and 8% were other coarse agricultural products.[38] Only in the 1970s, after the successful switch from dry land to paddy, did per capita grain production start to increase quickly. By 1975, rice paddy fields had increased to almost 500,000 *mu*, and rice accounted for 30% of all agricultural produce; the food supply in Lianshui had improved. By the 1980s, it had reached the average per capita grain production level for Jiangsu Province.

Table 4.2 Lianshui County population and food production

Year	*Population (10000)*	*Total grain output (10,000kg)*	*Per capita food production (kg/ person)*	*Paddy fields (10,000 mu)*	*Rice production*
1950	46.20	7744.2	167.6	0	0
1955	52.60	16722.4	317.9	0	0
1960	55.83	10592.1	189.7	19.12	1012.1
1965	60.39	11434.1	189.3	0.19	23.0
1970	72.59	18718.9	257.9	7.83	1214.1
1975	78.81	26181.6	332.2	46.79	7408.6
1980	82.17	35227.7	428.7	42.1	11810.2
1985	86.65	59585.5	687.7	61.92	30447.9

Source: Edited by Lianshui County Gazetteer Editorial Committee, Lianshui County Gazetteer, *Jiangsu Ancient Books Publishing House*, 1997: population statistics, Table 4.1 (p. 171), food production statistics, Table 7.5 (pp. 244-245)

In addition to the insufficient supply of grains, meat products were also scarce. Residents' main sources of animal protein were pork, beef, lamb, poultry, eggs, and fish. According to one investigation, due to poverty, before

36 Lianshui County Water Conservancy Gazetteer Editorial Committee, *Lianshui County Water Conservancy Gazetteer*, pp. 81-82.

37 Ibid., p. 64.

38 Lianshui County Gazetteer Editorial Committee, *Lianshui County Gazetteer*, pp. 244-245, Table 7.5.

China's economic reform, ordinary villagers could only afford to buy a little pork at Chinese New Year (consumption of beef and mutton was minimal), and due to a lack of 'fat' in everyday life, 'fatty' pork was given priority. Of course, during that period, it was common everywhere in rural China for people to eat only a little pork, beef, and mutton. Even people in relatively prosperous villages in areas like Sunan (southern Jiangsu) also ate very little meat. But with regard to aquatic products, Lianshui County was much worse than the average situation in Jiangsu province. As shown in the table below, Lianshui County's per capita aquatic product production was only around 1/10 of the provincial level in the 30 years before the reform.

Table 4.3 Comparison of per capita main agricultural production: Jiangsu Province and Lianshui County

Year	*Lianshui County Population (10000)*	*Jiangsu Grain Yield (kg/ person)*	*Lianshui County Grain Yield (kg/person)*	*Jiangsu Pigs (head/ person)*	*Lianshui Pigs (head/ person)*	*Jiangsu Province Aquatic Products Per Capita (kg/ person)*	*Lianshui County Aquatic Products Per capita (kg/ person)*
1950	46.20		167.6		0.33		
1952	48.70	270.0	260.8	0.24	0.28	4.55	
1955	52.60		317.91		0.44		0.078
1957	54.12	257.0	200.0	0.34	0.14	6.90	0.18
1960	55.83		189.72		0.27		1.00
1962	57.16	225.0	127.2	0.22	0.19	4.51	0.87
1965	60.39	316.0	189.3	0.46	0.41	5.50	0.64
1970	72.59	329.0	257.9	0.50	0.41	5.15	0.20
1975	78.81	367.0	332.2	0.60	0.46	6.45	0.53
1980	82.17	408.0	428.7	0.70	0.55	7.22	1.03
1985	86.65	505.0	687.7	0.64	0.68	10.91	2.31

Source: Jiangsu Province 1994 Statistical Yearbook, p. 138; Lianshui County Gazetteer Editorial Committee, Lianshui County Gazetteer, *Jiangsu Ancient Books Publishing House*, 1997, pp. 171, 271-272, 244-245, 271, 272, 276-277

Water Control, Crop Changes and Drinking Water Source Pollution

After the Yellow River overflowed into the Huai, disastrous floods in Lianshui caused poverty and suffering. Since the establishment of the PRC in 1949, almost half a century of water conservancy and governance has almost completely resolved the problem of flooding and also brought drought under control. The switch from dry land to paddy has also more or less eliminated the problem of salinization. However, this process of water management and changes in land use has also had a profound effect on farmers' living environment and unintentionally created new kinds of health risks.

The first goal of water conservancy in Lianshui was to solve the problem of how to discharge stagnant water. To do this, and at the same time increase agricultural production, an irrigation system was rapidly constructed. The other major issue was soil salinization that had resulted from the Yellow River's overflowing into the Huai. Converting dry land to paddy, in combination with irrigation and other agricultural works, completely resolved this problem. Using water to wash off and suppress salt not only solved the problem of salinized wasteland but also produced high yields of good quality rice. This process began as an experiment in the 1950s, and quickly achieved results.

> In the salinized wasteland where nothing could be planted before, rice production is already 400 to 500 *jin* (half kilo) in the first year. For example, there is a piece of saline-alkaline waste land in the Chengbei Group of Dongfeng Commune, which had never been used for crops. In 1959, it was planted with rice, and its seasonal average production is now 488 *jin* per *mu*.[39]

Lianshui County statistics show the progress of the changes. From 1956, rice was planted on an acreage of 13,200 *mu*. This increased to almost 200,000 *mu* in 1959 and 1960 but decreased to less than 10,000 *mu* in the following few years. However, the real breakthrough came in the 1970s. In 1971, rice covered 78,300 *mu*, and by 1971, it was more than 200,000 *mu*. By 1980, it had doubled to 421,000 *mu*, and by 1985 it was up to 619,200 *mu*, remaining stable thereafter. After converting dry land to paddy, land that was planted with corn and potatoes in the past was used to plant good quality, high yield agricultural crops. This resulted in the output of thousands of *jin* of grains and rice per *mu* compared with hundreds of *jin* of coarse grains in the old days. This was a historic change for people who never had enough to eat in the past. In 1983 when the Land Responsibility System started, rice had already become the main agricultural product of Lianshui County, which had been transformed from a poor county known for famine into a major site of grain production.[40]

39 Lianshui County Water Conservancy Gazetteer Editorial Committee, *Lianshui County Water Conservancy Gazetteer*, p. 204.

40 Lianshui County Gazetteer Editorial Committee, *Lianshui County Gazetteer*, pp. 244-245.

However, converting dry land to paddy also changed the water system, including drinking water sources. It connected farmers' ponds with the discharge system of the agricultural fields, enabling fertilizers, pesticides, as well as excess manure, to enter the drinking water supply and increasing health risks. Below I explain these changes in detail, based on fieldwork in Bei Village of Lianshui.

The land in Lianshui is very flat; with the exception of the area along the Yellow River, which is slightly higher, the other areas are comparatively low-lying. To prevent flooding, farmers would build a 'village platform' (*zhuangtai*) before they built their homes and they would construct a rectangular pond about 40 to 50 metres long in front of it. The width of the pond ranges from about 30 to 50 metres, and the length is mostly the same as the length of the house. For row houses, the pond is long. Soil from the pond was used to increase the height of the foundation of the *zhuangtai*, which rise about one metre above the surrounding land, and the area that was dug out naturally filled with water and became a pond (see Figure 4.3 below). There are two major water sources for these ponds: rainwater accumulated from around the pond and groundwater that seeps in from below. In fact, besides providing soil and sand to increase the height of the *zhuangtai*, the pond also plays a very important role in village life, functioning in a similar way to rivers or streams in southern China, and providing water for drinking or washing clothes. Because it acts as a source of drinking water, the quality of the pond water is closely related to villagers' health.

In Bei Village, some households live in clusters, which are called 'Zhuang'. There are a total of thirteen Zhuang in Bei Village, and one Zhuang constitutes a Village Group (similar to a Production Team during the People's Commune period). There is usually one pond per Zhuang, and these are called 'river ponds' (*hetang*) or 'big wells' (*dajing*) by the locals. If the pond is to the south of the house, then unclean places such as the toilet are at the back of the house. If there are more than two rows of houses in one Zhuang, then the pond is in between them, and there is a certain distance between the houses and the pond. Judging from what I saw, that distance is about 40 metres. This layout makes it convenient for villagers to use the water, but it also means the pond is far enough away from the activities of daily life to keep it fairly clean.

Unlike quays beside rivers or streams and ponds in mountainous area, the pond in front of the *zhuangtai* has special characteristics. It is easy to understand why villagers call it a 'river pond' because it looks like a river, only shorter. But why is it also called a 'big well'? This is because it works in a similar way to a well: due to the rich groundwater in the area, digging

Figure 4.3 Village and water layout in the Bei Village area of Lianshui County

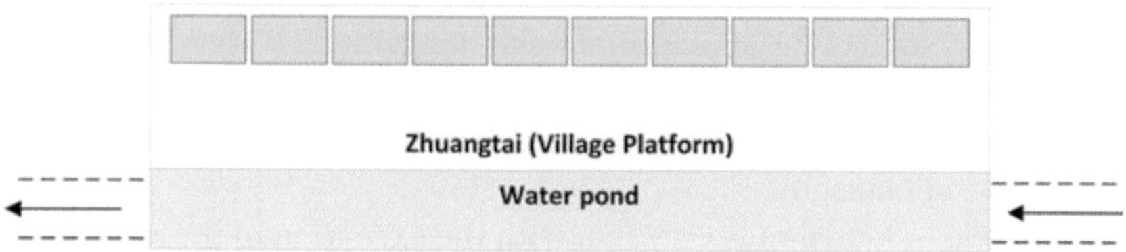

anywhere will easily draw up groundwater, and it will not dry up. However ordinary wells are no bigger than one square metre, and the surface of the pond covers several *mu*, so it is called a 'big well'.

Practices also vary in different parts of the same county. In November 2014, during the Beijing Environment and Health Annual Conference, I discussed this issue with Mr. Shi, who has been working in environmental protection for years. When he was a child, Mr. Shi lived for a dozen years in Huidun in northern Lianshui County. In the villages he is familiar with, there are also ponds that are formed the same way as in Bei Village, but they are called 'big ponds' (*wangtang*) by the locals. These big ponds are more like an 'oxygenized pond' with reeds, weeds, lotus, and water grass growing in and around them. A limited amount of fish and shrimps are also farmed in these ponds, and the villagers do their washing and swim in them. Because they are relatively large, human activities do not affect the water quality.

Regardless of whether they are 'river ponds', 'big wells' or 'big ponds', before the dry land was converted to paddy, these ponds were all independent water bodies separate from the external water system. In terms of pollution-related health risks, there were two possibilities. The first was small ponds accumulating a high density of sewage. One source of water for this type of pond is rain water that accumulates on the surrounding land, and pollutants could be washed into the pond along with the rainwater. This kind of pond is similar to the 'dry wells' (*hanjing*) in Lin County;[41] and

41 'In most places of Lin County, you can dig underground for hundreds of metres and not find water. Therefore, people dig ponds to store rainwater. This type of pond is called a 'dry well'. In 1949, there were about 10,600 dry wells, [...] by 1957 there were 27,120 [...] By 1965, this had increased to 38,895. [...] Between 1981 and 1985, a total of 34,222 new dry wells were dug. In Zhang Village of Haodi [...] there is one well per person on average. This not only guarantees drinking water for people and domestic animals, but also solves the problem of the lack of irrigation water in the spring planting season'. Lin County Gazetteer Editing Committee, *Lin County Gazetteer*, pp. 242-243. People who use dry wells have a high incidence of oesophageal cancer, with an onset rate of 112.90/100,000 and a mortality rate of 89.86/100,000. People who use tap water have an onset and mortality rate of 57.88/100,000 and 54.32/100,000, respectively. See,

if used for drinking water, this could present a quite serious health risk. 'Big ponds', which have a larger area and a natural complex ecosystem have the capacity to dilute or dissolve external pollutants that may enter them. The water therefore remains clean and there is only a very small health risk.

After the dry land was converted to paddy fields, with the exception of the *zhuangtai*, all the other land in the village was used to plant rice. Rice needs to be well irrigated, and this requires a relatively comprehensive irrigation and drainage system. During the construction of this system, the ponds, which were originally separated, were intentionally or unintentionally connected to the irrigation canals. As shown in Figure 4.3, after these changes, water ponds in the village become part of the agricultural irrigation and drainage system. This change created new health risks because it made the water quality in the ponds dependent on the water quality in the irrigation system. During the rice growing seasons of summer and fall, the paddies need frequent irrigation and draining and so the water quality of the pond is basically determined by the water quality in the paddy fields. This poses far more health risks than when the ponds were isolated from the agricultural water system.

Because rice is prone to pest infestations, large quantities of pesticides are used in its cultivation. To maintain high yields, since 1970s, large quantities of fertilizer have also been used on paddy fields. In 1953, Lianshui County used 5 tons of fertilizer, in 1960, it was 1,086 tons, in 1970, 6,035 tons, in 1980, 59,827 tons and in 1987, 178,880 tons. 'The average amount of fertilizer used per *mu* has tended to increase, in 1982, it was 69kg, and in 1987, it was 134.9kg'.[42]

In addition, since the 1980s, livestock farming has developed rapidly in rural areas of Lianshui, and manure has increased as a consequence. The higher output of grain production laid the basis for rearing pigs and poultry, and these activities increased quickly after the implementation of reform and opening up in Lianshui. By 1987, there were 641,000 pigs, four times more than in 1954 when there were only 151,400. The average weight of the pigs sold was 124 kilograms, which was twice the average weight in 1950 of 60 kilograms (see Table 4.4 below). The county's pigs produced about eight times as much excrement in 1987 as they did in 1950.

The record shows that in 1977, poultry numbered 718,700. By 1980 this had increased to 1,020,000, and by 1985, to 7,331,700, or ten times as many

Cheng, Lian, Liu et al., 'The Impact of Water Change on Esophageal Cancer Onset and Mortality Rates', China Cancer, 2007, pp. 998-999.

42 Lianshui County Gazetteer Editing Committee, *Lianshui County Gazetteer*, p. 256.

birds as in 1977 (see the table below). If there are 180,000 poultry farming households in Lianshui, with an average of 40 birds per household, a village with 30 such households would have more than 1,200 fowl.

Manure produced by animals raised by villagers accumulates on farm land, and some of it will be washed from the paddy fields into the water ponds. Manure that is not absorbed by the rice will flow into the ponds and pose a health risks to the villagers' drinking water. Based on the 'Nitrogen Cycle Hypothesis' proposed by Xu Zhixia et. al, if the amides from farm manure enter the human body along with nitrates and nitrites, they will create highly carcinogenic nitrosamines. In this 'risk chain', manure entering the drinking water is a crucial link. This situation, in which a change in the crops farmed causes problems with drinking water quality, resembles what happened in the rural area of Lin County studied by Xu Zhixia and his colleagues, where there was a high incidence of oesophageal cancer. Talking to local medical service personnel in Lianshui during fieldwork, I found that although they did not know about the high incidence of oesophageal cancer and its causes in Lin Country, their presumption of the causes of cancer in their area was very similar. For example, Mr. Ni said that in his home village in Zhuma Township, an old couple got cancer and their water pond is very dirty. Zhuma Township is the place that first successfully converted its dry land to paddy. Two doctors, both with the family name Liu, in a new village near Bei Village that I visited also referred without hesitation to changes in the surrounding environment, for instance, the large amounts of pesticides and fertilizers used. This also resembles what Anna Lora-Wainwright found in her research in Sichuan Province, where a villager called Baohua insisted that the high incidence of cancer was related to well water pollution, 'the well water is not good, [the well is] too shallow, and right beside the pigpen, sewage ditch and paddy fields'.[43] The CDC conducted tests but did not detect any definite carcinogens, but in fact the situation is very similar to the drinking water pollution in Lin County studied by Xu Zhixiang et al.

Improving Drinking Water

Su Yang and Duan Xiaoli have divided environmental health problems in China into two periods. Before the 1980s, traditional health risks dominated,

43 Lora-Wainwright, 'Anthropological Study of Cancer Villages: Perception of Villagers on Accountability and Coping Strategies', p. 258.

Table 4.4 Fertilizer usage and livestock farming in Lianshui County

Year	*Total (tons)*		*Nitrogen fertilizer (tons)*		*Pigs (ten thousand)*	*Poultry (ten thousand)*
	Physical quantity	*Effective component*	*Physical quantity*	*Effective component*		
1953	5	5	5	5	20.50	–
1955	45	41	45	41	23.25	–
1960	1,086	931	1,086	931	15.11	–
1965	3,874	3,320	3,874	3,320	24.92	–
1970	6,035	4,976	2,632	2,256	29.62	–
1975	17,426	14,935	8,744	7,700	36.50	–
1980	59,827	54,440	36,895	33,330	45.12	102.00
1985	137,435	38,809	76,672	23,002	59.25	733.17
1986	155,333	44,024	88,580	26,572	65.03	632.00
1987	178,880	49,220	101,464	28,410	64.10	625.21

Data source: Lianshui County Gazetteer, edited by Lianshui County Gazetteer Editorial Committee, Jiangsu Ancient Books Publishing House, 1997 (pp. 256-257, 271-272)

particularly infectious diseases caused by bacteria and parasitic diseases due to the lack of good infrastructure and public services. These health problems can be solved through strengthening infrastructure, including improving water and toilet facilities, and providing medical services. After the 1980s, environmental health problems in China have gradually shifted to modern risks caused mostly by chemical pollution resulting from industrialization and urbanization. Su and Duan noted that China had yet to develop a mature system for preventing and controlling these risks and these systems lagged behind actual demand.[44] Looking at the situation in Lianshui and other places, where the construction of large scale irrigation systems resulted in changed drinking water sources and quality, there has been no clear, immediate or concrete responsive strategy, whether in theoretical research or government policy. Objectively, the situation is still constrained by the fact that technical awareness lags behind and there is a lack of economic resources to improve things.

Nonetheless, from a long-term perspective, drinking water in Lianshui has actually shown continuous improvement. If Lin County (Linzhou) in Henan Province represents a case of perception-response in which active research was carried out to decrease the incidence of oesophageal cancer, Lianshui in Jiangsu represents another type of case that has not

44 Su and Duan, 'The Status, Problems and Responses of Environment Health Work in China', pp. 73-74.

yet been listed as a place of concern. However, along with economic and social development, the rise in living standards, and improvements in public medical services, over the relatively long term from 1970s to the early twenty-first century, improvements to drinking water have been made continuously. Based on fieldwork and a review of the literature, improvements to the water supply in rural Lianshui can be divided into three stages.

The first stage was government-initiated point by point improvement. After the establishment of the new China, the environment in towns and villages was often said to be 'dirty, messy and bad' (*zangluancha*), with infectious diseases such as cholera, haemorrhagic fever, and plague rampant. Flies, mosquitos and mice were the major carriers of these diseases. The main aim of the Patriotic Hygiene Campaign was to eliminate these 'pests' and break the links that allowed these diseases to be transmitted. Because rural areas had bad hygiene conditions, the government used the 'Patriotic Hygiene Campaign' to improve the situation through measures such as changing the water supply, changing toilet facilities, household waste management, and the prevention and control of disease through biological means, promoting health education, environmental hygiene management, etc. The hygiene department sent personnel to the countryside to conduct health propaganda and education such as urging people to wash their hands before meals or after the toilet, change their clothes more often and not drink water without boiling it first.[45]

The second stage started in the 1970s based on the needs of the 'Patriotic Hygiene Campaign' related to environment and health and it took the form of a combination of government water supply initiatives and villagers' own efforts. In 1977, a tap water plant was set up in Lianshui. By 1980, two small water towers had been built in Chengji Commune, together with 1,574 small wells, benefiting 18,030 people or 85% of the total population in the area. The changing of the water supply had an obvious effect in preventing intestinal infectious diseases. In 1979 there were 4,893 cases of intestinal infectious diseases (diarrhoea, enteritis etc.) By 1980, the number had decreased to 1,261, or about a quarter of the previous level. There were 221 cases of contagious hepatitis in 1979, and this decreased to 51 in 1980, a reduction of 93.2%.[46] By the end of 1987, simple tap water plants had been built in 19 township government seats, benefiting about 27,000 residents, or 4.5% of the rural population. Villagers also built 38,400 small wells with

45 Lianshui County Gazetteer Editorial Committee, *Lianshui County gazetteer*, p. 827.
46 Lianshui County Gazetteer Editorial Committee, *Lianshui County Gazetteer*, p. 830.

hand-pumps, enabling more than 245,000 people to use well water, or more than 30% of the total rural population.[47] From my fieldwork I found that families in Bei Village who had relatively good economic status and who were also comparatively informed about modern technology had all started to use small hand-pump wells by that time. For example, this was the case for Mr. Wang's family and other households.

The third stage was building deep wells. In Qianji Township in Lianshui, the construction of deep wells went through two phases. The first was is in 1998, when seven water plants were built to solve the drinking water problem for eight administrative villages (including townships) with a population of 22,500. Eight years later in 2006, three more water plants were built to cover almost 10,000 people in four administrative villages. By 2006, then, all villages in Qianji Township were using deep well water.

Table 4.5 Basic situation of water plants in Qianji Township, Lianshui County

Water Plant Name	*Depth (m)*	*Date established (year, month)*	*User Population*	*Area*
Liuqiao Water Plant	150	2006.07	1974	Liuqiao
Jincheng Water Plant	198	2006.07	4249	Jincheng, Kongwang
Xindengta Water Plant	180	2006.07	3246	Xindengta
Hewan Water Plant	182	1998.08	4171	Hewan, Xialou
Panma Water Plant	180	1998.10	4361	Panma
Wangliu Water Plant	160	1998.10	4622	Wangliu
Yanshou Water Plant	174	1998.10	1619	Yanshou
Liyuan Water Plant	186	1998.11	1815	Liyuan
Qianji Water Plant(in the township government)	174	1998.11	2672	Qianji Township
Lizhuang Water Plant	128	1998.11	3251	Lizhuang

Source: Lianshui County Qianji Township Disease Prevention and Healthcare Office

In summary, the logic behind changing the water supply in rural Lianshui was mostly to improve hygiene and health, but it was not specifically targeted at oesophageal cancer. This is in keeping with the general logic of local economic and social development. The 'Patriotic Hygiene Campaign' was an important responsibility of the local government and decreasing disease, improving the state of public health and reducing common intestinal illnesses through this campaign was a major part of the government's routine work.

47 Ibid.

From villagers' point of view, to the extent that they could afford it, clean water was obviously desirable because it was more sanitary and hygienic. As economic and living conditions improved, technology progressed, and traditional drinking water sources deteriorated. Even though they could not precisely understand the relationship between water and diseases such as oesophageal cancer, changing water sources was one effective measure that villagers could take to avoid health risks.

Conclusion

The upper reaches of the Yellow River were once well-covered with vegetation but with the increase in human activities, this gradually became a barren area with severe soil erosion. Large quantities of silt flow down the Yellow River to its middle and lower reaches, making the river bed higher and higher, and resulting in continuous breaches and flooding. The overflow from the Yellow River took over the channel of the Huai, causing the most severe and long-lasting environmental disaster in Chinese history. Due to its geographical location, Lianshui County experienced repeated flooding, as well as secondary disasters such as drought caused by the chaotic river system, and soil salinization. Disasters also created poverty, famine, imbalanced diets and malnutrition, resulting in health risks.

To put an end to the flooding and consequent disasters, and solve the livelihood problem, large scale water conservancy projects were started in the Lianshui area right after the establishment of the People's Republic of China. Afterwards, dry land was also successfully converted to paddy fields. Within 30 years, flooding and drought were eliminated and people's livelihoods were secured.

However, the change in agricultural crops also unintentionally changed the water system, water sources and water quality, causing new health risks. From a historical perspective, we can now see that the change of drinking water sources due to the need for irrigation has been a common phenomenon. Research has shown that the high incidence of cancer in Yangzhong County in Jiangsu Province is related to the irrigation system there which leaves the ditch water stagnant or half stagnant. 'The paddy fields are closed for nine months, and so people drink the water from the paddy fields during these months'.[48] The situation is similar in Jianhu

48 See Xu, ed., *Fertilizer, Waste Water and Oesophageal Cancer*, p. 77.

County near Lianshui. 'Ditches and canals are often near residents' houses. Drinking water is from surface water such as rivers, canals etc. The water is murky. Every rainy season, household waste water and water from the paddy fields after fertilizing all runs into the ditches, making them seriously polluted'.[49]

From the perspective of the environment-health relationship, a problem that looks very clear later is not always clear-cut while it is happening. For instance, in Lin County in Henan, and Qidong City in Jiangsu, the high incidence of oesophageal cancer and liver cancer was already discovered during the 1950s and 1960s, and attracted a lot of attention, resulting in the establishment of the death registration system and local cancer control and prevention research. Measures to address the high incidence of cancer have been explored for a long time, but not all places attract the attention of the state as Lin County, Yanting and Qidong did. For example, the high incidence of oesophageal cancer in northern Jiangsu was not effectively addressed for a long time. However, along with the development of the health and welfare system, including improving drinking water, the risk of oesophageal cancer has declined.

During the past half century, and especially the past 30 years, China has been experiencing a rapid transformation of its natural environment, economy, society and culture. It is likely that over the coming 20 to 30 years, or even longer, China will continue to develop and change rapidly. Some changes in the environment are likely to attract the attention of the government and the public, and receive a relatively quick response, but other changes will be slower and some health problems may already be very severe by the time they become visible.

49 Ibid., p. 104.

Appendix

Table 4.6 List of Bei Village malignant tumour patients

Serial No.	*Name*	*Gender*	*Year of birth*	*Year of disease onset*	*Type of cancer*	*Alive/decease time*	*Relatives disease situation*	*Village*
1	XDH	Male	1922	1984	Oesophageal cancer	1994		Kongzhuang
2	WJT	Male	1923	1990	Liver cancer	1991		Wangzhuang
3	WHB	Male	1916	1990	Oesophageal cancer	1992		Wangzhuang
4	WXK	Male	1929	1990	Oesophageal cancer	1993		Hezhuang
5	WLQ	Male	1921	1991	Oesophageal cancer	1994		Wangzhuang
6	WZS	Female	1921	1992	Oesophageal cancer	1995	Mother of NO7 WSB	Hezhuang
7	WSB	Male	1938	1993	Oesophageal cancer	1994		Hezhuang
8	HZS	Female	1930	1993	Liver cancer	1996	Wife of NO11 HXJ	Kongzhuang
9	HXH	Male	1946	1994	Oesophageal cancer	1996		Hezhuang
10	KXX	Male	1934	1994	Oesophageal cancer	1997		Kongzhuang
11	HXJ	Male	1932	1994	Oesophageal cancer	1997		Kongzhuang
12	YZD	Male	1944	1994 2002	Lung cancer Oesophageal cancer	2004		Lizhuang
13	WRT	Male	1926	1996	Intestinal cancer	1999		Wangzhuang
14	ZXY	Female	1940	1997	Oesophageal cancer	1998		Xiaodong
15	WHT	Male	1952	1997	Oesophageal cancer	Alive		Dongyang
16	LRJ	Male	1937	1997	Oesophageal cancer	Alive		Kongzhuang
17	DHB	Male	1957	1997	Oesophageal cancer	Alive		Dongyang
18	WXJ	Male	1937	1997	Oesophageal cancer	Alive		Wangzhuang
19	HXF	Male	1930	1997	Oesophageal cancer	Alive		Lizhuang
20	JXT	Male	1939	1998	Oesophageal cancer	1999		Laozhuang

Serial No.	*Name*	*Gender*	*Year of birth*	*Year of disease onset*	*Type of cancer*	*Alive/decease time*	*Relatives disease situation*	*Village*
21	WXF	Male	1932	1998	Oesophageal cancer	2000		Sunzhuang
22	WXZ	Male	1938	1998	Oesophageal cancer	2000		Xiaodong
23	WLB	Male	1943	1999	Oesophageal cancer	2000		Mawang
24	HZM	Male	1941	1999	Oesophageal cancer	2001		Hezhuang
25	HWS	Female	1931	1999	Oesophageal cancer	2001	Mother of NO9 HXH	Hezhuang
26	HXW	Male	1930	1999	Lung cancer	2001		Kongzhuang
27	WRQ	Male	1949	1999	Oesophageal cancer	2001		Xiaodong
28	WHQ	Male	1937	1999	Oesophageal cancer	2002		Xiaodong
29	LZB	Male	1938	1999	Oesophageal cancer	2003		Laozhuang
30	WXF	Female	1949	1999	Oesophageal cancer	2006		Lizhuang
31	WZS	Male	1943	1999	Oesophageal cancer	Alive		Wangzhuang
32	WFY	Female	1943	2001	Oesophageal cancer/rectal cancer	Alive		Qianzhuang
33	WLB	Male	1944	2002	Lung cancer	2004		Wangzhuang
34	YQ	Male	1941	2002	Oesophageal cancer	2007		Xiaodong
35	XLF	Female	1945	2002	Oesophageal cancer	Alive		Dongyang
36	WYB	Male	1945	2004	Lung cancer	2005		Mawang
37	WGM	Male	1943	2004	Lung cancer	2006		Dongyang
38	LSN	Female	1939	2004	Oesophageal cancer	2009		Xiaodong
39	ZXF	Female	1934	2004	Cardiac cancer	Alive		Lizhuang
40	LCJ	Male	1932	2005	Cardiac cancer	Alive		Lizhuang
41	LGZ	Male	1941	2005	Oesophageal cancer	Alive		Qianliu
42	ZSZ	Male	1941	2005	Rectal cancer	2009		Sunzhuang
43	MYJ	Male	1957	2005	Oesophageal cancer	2013		Mawang

Serial No.	*Name*	*Gender*	*Year of birth*	*Year of disease onset*	*Type of cancer*	*Alive/decease time*	*Relatives disease situation*	*Village*
44	MYF	Male	1948	2006	Lung cancer	2009		Mawang
45	WCQ	Male	1951	2006	Cardiac cancer	2012		Laozhuang
46	LGJ	Male	1958	2006	Oesophageal cancer	Alive		Houliu
47	LYJ	Male	1943	2007	Oesophageal cancer	Alive		Houliu
48	ZCL	Male	1942	2007	Oesophageal cancer	Alive		Houliu
49	WGB	Male	1955	2007	Liver cancer	2008		Wangzhuang
50	ZJZ	Male	1930	2008	Oesophageal cancer	2009		Dadong
51	XYM	Female	1938	2008	Oesophageal cancer	Alive		Wangzhuang
52	LSZ	Female	1961	2008	Rectal cancer	Alive		Sunzhuang
53	WYB	Male	1941	2009	Lung cancer	Alive		Hezhuang
54	YZT	Male	1940	2009	Lung cancer	Alive		Lizhuang
55	DGH	Male	1960	2009	Oesophageal cancer	Alive		Dongyang
56	ZSH	Male	1951	2009	Oesophageal cancer	Alive		Mawang
57	WXC	Male	1941	2009	Oesophageal cancer	Alive		Qianzhuang
58	LHJ	Male	1955	2009	Stomach cancer	Alive		Qianliu
59	LCH	Female	1929	2009	Liver cancer	2009		Lizhuang
60	ZSC	Male	1946	2009	Liver cancer	2009		
61	WXF	Female	1929	2009	Oesophageal cancer	2011		Lizhuang
62	WYL	Female	1936	2009	Lung cancer	2011		Mawang
63	ZMY	Female	1930	2010	Oesophageal cancer	2010		Houliu
64	KQY	Male	1945	2010	Oesophageal cancer	2011		Kongzhuang
65	LYS	Male	1941	2010	Oesophageal cancer	2011		Qianliu
66	MYF	Female	1949	2010	Lung cancer	2012		Mawang
67	WGT	Male	1950	2010	Lung cancer	2012		Dongyang

Serial No.	*Name*	*Gender*	*Year of birth*	*Year of disease onset*	*Type of cancer*	*Alive/decease time*	*Relatives disease situation*	*Village*
68	LYS	Male	1934	2010	Oesophageal cancer	2013		Qianliu
69	ZHQ	Male	1970	2010	Rectal cancer	Alive		Dadong
70	WEL	Male	1952	2010	Oesophageal cancer	Alive		Dongyang
71	WXC	Male	1938	2010	Bladder cancer	Alive		Xiaodong
72	XYQ	Male	1955	2010	Brain tumour	Alive		Kongzhuang
73	ZYM	Female	1984	2010	Breast cancer	Alive		Lizhuang
74	HZB	Male	1933	2011	Rectal cancer	Alive		Hezhuang
75	LZJ	Male	1939	2011	Oesophageal cancer	Alive		Qianliu
76	YYQ	Male	1958	2011	Cardiac cancer	Alive		Lizhuang
77	BLY	Female	1954	2011	Lung cancer	Alive		Dongyang
78	LR	Male	1941	2011	Oesophageal cancer	2011		Qianliu
79	WNQ	Male	1965	2011	Lung cancer	2011		Laozhuang
80	LKS	Male	1931	2011	Oesophageal cancer	2011		Houliu
81	WXM	Male	1927	2011	Lung cancer	2012		Dongyang
82	WWT	Male	1953	2011	Lung cancer	2013		Hezhuang
83	WSL	Male	1945	2012	Lung cancer	2012		Qianzhuang
84	WYL	Male	1941	2012	Lung cancer	2012		Qianzhuang
85	WTQ	Male	1948	2012	Lung cancer	2013		Laozhuang
86	DYZ	Female	1934	2012	Oesophageal cancer	2013		Hezhuang
87	WXC	Male	1933	2012	Lung cancer	Alive		Hezhuang
88	WLF	Female	1941	2012	Oesophageal cancer	Alive		Qianzhuang
89	ZHC	Male	1970	2013	Oesophageal cancer	alive		Dadong
90	WXY	Male	1948	2013	Laryngocarcinoma	alive		Laozhuang
91	LSL	Female	1936	2013	Oesophageal cancer	2013		Xiaodong

Works Cited

Balshem, Martha. *Cancer in the Community.* Washington and London: Smithsonian Institution Press, 1993.

Chen, Ajiang. 'Inside and Outside "Cancer Village"'. *Journal of Guangxi University for Nationalities, Philosophy and Social Science Edition* 2013 (issue 2): 74-80.

Cheng, Lanping, Lian Shiyong, Liu Zhicai, et al. 'Influence of Water Quality Improvement on Incidence and Mortality with Esophageal Cancer in Linzhou, Henan'. *Bulletin of Chinese Cancer* 2007 (issue 12): 998-999.

—. 'The Impact of Water Change on Esophageal Cancer Onset and Mortality Rate'. *China Cancer* 2007 (issue 12): 998-999.

Elvin, Mark. *The Retreat of the Elephants – China's Environmental History.* New Haven and London: Yale University Press, 2004.

Gao, Shugeng, ed. *Experts Talk on Countering Oesophageal Cancer.* Beijing: Peking Union Medical College Press, 2014.

Han, Jianying, Lian Shiyong, Xu Zhixiang, et al. 'Influence of Standardized Deep Water on Morbidity and Mortality of Esophageal Cancer in Linzhou'. *Journal of Environment and Health* 2005 (issue 3): 200-202.

Hao, Xishan, and Wei, Yuquan, eds. *Oncology.* Beijing: People's Medical Publishing House, 2014.

Li, Qianwen, Yuan, Guangjin, Du, Yuxiang, et al. 'Analysis Research of Huai'an Region Oesophageal Cancer Epidemic Characteristics and Treatment Situation'. *Chinese Clinical Oncology* 2012 (issue 2): 142-145.

Lianshui County Gazetteer Editorial Committee. *Lianshui County Gazetteer.* Nanjing: Jiangsu Ancient Books Publishing, 1997.

Lianshui County Water Conservancy Gazetteer Editorial Committee. *Lianshui County Water Conservancy Gazetteer.* Changchun: Jilin Literature and History Publishing House, 2003.

Lin County Gazetteer Editing Committee. *Lin County Gazetteer.* Zhengzhou: Henan People's Publishing House, 1989.

Lora-Wainwright, Anna. 'Anthropological Study of Cancer Villages: Perception of Villagers on Accountability and Coping Strategies'. In Jennifer Holdaway, Wang Wuyi, Ye Jingzhong et al. (eds.), *Environment and Health: An Interdisciplinary View,* 238-262. Beijing: Social Sciences Academic Press, 2010.

Ma, Libo (Robert Marks). *Tigers, Rice, Silk, and Silt: Environment and Economy in Late Imperial South China.* Translated by Wang Yuru and Guan Yongqiang. Nanjing: Jiangsu People's Publishing House, 2011.

Ministry of Health, Office of Cancer Prevention and Treatment, ed. *China Malignant Tumor Mortality Investigation.* Beijing: People's Medical Publishing House, 1980.

National Tumour Prevention Office. *China Malignant Tumor Mortality Investigation (1990-1992)*. Beijing: People's Medical Publishing House, 2008.

Su, Yang, and Duan Xiaoli. 'The Status, Problems and Responses of Environment Health Work in China'. In Jennifer Holdaway, Wang Wuyi, Ye Jingzhong et al. (eds.), *Environment and Health: An Interdisciplinary View*, 73-74. Beijing: Social Sciences Academic Press, 2010.

Sun, Weixin. 'Analysis of 2013 Lianshui County Residents Malignant Tumor Deaths'. *Jiangsu Health Care* 2014 (issue 3): 34-35.

—. 'Characteristics of 2001 High Incidence of Lianshui County Resident Malignant Tumor'. *Jiangsu Journal of Preventative Medicine* 5 (2015): 43-44.

Wang, Suping, ed. *Epidemiology*. Beijing: Peking Union Medical College Press, 2009.

Xu, Zhixiang, ed. *Fertilizer, Waste Water and Esophageal Cancer*. Beijing: Science Publishing House, 2003.

—, Tan Jiaju, Chen Fenglan, et al. 'Farm Compost polluted Water May Induce Pharyngo-Esophageal, Gastric and Liver Carcinoma in Chicken'. *Chinese Journal of Oncology* 2003 (issue 4): 344-347.

—, Tan Jiaju, Han Jianying, and Chen Fenglan. 'Esophageal Cancer, Stomach Cancer, Liver Cancer Nitrogen Cycle Hypothesis and Testing'. *Journal of Research in Medical Sciences* 2008 (issue 1): 5-7.

Zhang, Daqing, ed. *History of Medicine*, Beijing: Peking University Medical Press, 2013.

Photograph 1 A secondary tributary of the Huai River upstream from Huangmengying Village (April 2009)

Photograph 2 Shaying River (April 2009)

Photograph 3 'Internal Pollution' of ponds in Huangmengying (April 2012)

Photograph 4 A new deep well in Huangmengying April (2012)

Photograph 5 A new image for the village, promoting the 'Circular Economy' (April 2009)

Photograph 6 Huangmengying named 'A Provincial Level Ecological Village' (August 2011)

Photograph 7	Subei Main irrigation channel (August 2007)
Photograph 8	The Julong chemical factory in the fields of Dongjing (January 2008)
Photograph 9	Pollution from the Julong chemical plant is emitted into local fields through the irrigation system (August 2007)
Photograph 10	Water from the Julong chemical plant enters the local production river of the village (January 2008)
Photograph 11	A typical spot for local water use in Dongjing (August 2007)
Photograph 12	Ducks reappear in the rivers around Dongjing after the Julong chemical plant is relocated (October 2011)

Photograph 13 The metal furniture factory in Xiqiao Village (May 2008)
Photograph 14 The subsidiary of the metal furniture factory located amid the paddy fields (December 2010)
Photograph 15 White waste water from the textile factories in Xiqiao (February 2007)
Photograph 16 Pollution on the surface of rivers in Xiqiao (May 2008)
Photograph 17 The chimney of the glass factory, photographed from a villager's home (December 2010)

Photograph 18 The new waste water plant being built by B Mine upstream of Liangqiao village (December 2010)

Photograph 19 The water quality in the Li River upstream from Liangqiao (December 2010)

Photograph 20 Steam from the chimney of the L Waste incineration power plant (December 2010)

Photograph 21 The online monitoring system in front of L Waste incineration power plant (December 2010)

Photograph 22 Fly ash solidification area of L Waste incineration power plant (December 2010)

5 A Prosperous 'Cancer Village'

Cheng Pengli

Abstract

Money, or life? Town and Village Enterprises helped local farmers get rich quickly, but they also caused serious environmental pollution that was a threat to people's health. Through a study of a village in northern Zhejiang, this chapter shows how, through their daily experiences, villagers established a logical relationship between industrial pollution and cancer and came to believe that the incidence of cancer was not only high, but strongly related to pollution. Facing health risks from pollution, the villagers initially engaged in resistance, but when this strategy was unsuccessful, they later switched to less active measures.

Keywords: Jiangnan, rural industrial pollution, risk perception, environmental resistance, village doctors

Lake Tai is situated in the Yangtze River Delta, a region with some of the strongest potential for development in China. Over the last 30 years, economic reform and modernization have put the region at the forefront of the whole country. Its impressive growth has stimulated the development of eastern China and the Yangtze River basin and contributed to the economic and cultural development of the country. A special feature of the region's economic development has been Town and Village Enterprises (TVEs), especially in the Southern Jiangsu (*Sunan*) and Northern Zhejiang (*Zhebei*) regions around Lake Tai. These have greatly increased the income and standard of living of rural residents; compared to rural parts of other regions in China, the area around Lake Tai has become prosperous, and its farmers rich.

Industrial development in villages and towns surrounding Lake Tai made farmers wealthy, but it also polluted the water and caused environmental degradation. In the 30 years after reform and opening up, industrial

Chen Ajiang, Cheng Pengli, and Luo Yajuan, *Chinese "Cancer Villages": Rural Development, Environmental Change and Public Health*. Taylor & Francis Group 2020
DOI: 10.5117/9789089647221_CH05

enterprises in the Lake Tai countryside grew in both size and number, and pollution emissions increased as a result. Due to the widespread coverage in the media, public attention to water pollution in Lake Tai has focused on the annual outbreaks of green-blue algae each summer, but in fact, water pollution in the region in recent years has had two characteristics. First, it is worse downstream from Lake Tai than in the lake itself, but because Lake Tai is the source of drinking water for the surrounding cities, more attention has been paid to controlling pollution in the lake. Second, water pollution in the valley is worse in rural areas with TVEs than in urban areas. However, due to the longstanding rural-urban dichotomy, water pollution in the countryside has never received effective control, while environmental remediation in the cities of the Lake Tai Basin in recent years has greatly improved environmental quality there, including water quality.

So, money or life? In the rural areas around Lake Tai, industrial development increased the incomes and quality of life of the average citizen, but the quality of their living environment deteriorated significantly, and pollution is threatening their health. The final goal of economic development is to promote overall development for the people. But if economic growth comes at the expense of people's health, then what meaning does this development have?

Background to the Case

Regional Background: geographic and socio-economic characteristics

The village in this case study – Xiqiao Village – is situated in the northwest suburbs of Jiaxing City, Zhejiang Province. Jiaxing is on the southeast shore of Lake Tai and is part of the core area of the Lake Tai Basin. An important geographical characteristic of the region is its dense river network and highly developed water system. The terrain of the Lake Tai Basin is asymmetrical, with a mountainous region to the west and the Lake Tai plain to the east. Mountains cover approximately 20% of the area, while the rest is flat. The Lake Tai plain is on average roughly ten meters above sea level, while the middle of the lake and its southeast shore are situated at the lowest altitude in the region. Water covers 5,551 square kilometres, or approximately 15% of the basin, including 3,159 square kilometres of lakes and 2,392 square kilometres of rivers. The rivers in the region are long and dense, and the numerous lakes and rivers criss-crossing the terrain have made the region famous for its water networks.[1]

Another characteristic of the Lake Tai Basin region is the polder system that has been formed over the course of history. A polder system is a man-made ecosystem formed by changing the natural environment and in the Tai Basin it has played an important role in the development of agriculture in the region.[2] The development of polders greatly increased the amount of arable land by transforming river canals, lakes and coastal shallows into fertile paddy fields and dry land which were then used to grow high-yield crops, resulting in increases in agricultural output. Over many years, agricultural practices have developed in which fish are raised outside the polders, rice is grown within them, and mulberry trees are planted on top of them, achieving a balanced ecological system.

The Lake Tai Basin region is historically a land famous for its prosperity. It was one of China's earliest agricultural hubs, and famous for its crafts and culture. After economic reform, the economic and social development of the region took off, and it became one of the most economically developed regions, with the highest urban density in China. In 2005 for example, approximately 3.5% of China's population lived in the basin region and it contributed 18.6% of the industrial output of the whole country, 11.6% of the Gross National Product, and 22.1% of fiscal revenue.[3] It should be emphasized that an important characteristic of the economic and social development of the Lake Tai Basin region is its Town and Village Enterprises. In the early 2000s, rural industries accounted for 43% of the region's industrial output, making them the mainstay of the rural economy. At the same time, the urbanization rate in the Lake Tai Basin region was already very high, at 73%.[4]

Rapid economic growth in the region went hand in hand with a serious decline in regional environmental quality and, in particular, with severe water pollution. Due to its unique geography and climate, the Lake Tai Basin was originally the famous Jiangnan 'land of fish and rice'. Over the last 30 years, however, the rapid development of industry and increase in population density led to a severe shortage of clean water in the area. Water pollution has already restricted socio-economic development to some

1 Data source: Lake Tai website, http://www.tba.gov.cn/. Accessed 27 July 2009.

2 A polder system is comprised of a low-lying tract of land enclosed by dikes.

3 Data source: Lake Tai website, http://www.tba.gov.cn/. Accessed 27 Jul 2009.

4 Gao et al., 'Lake Tai Basin Economic Development and the Demand for River Resources', pp. 62-64.

extent and poses a serious threat to the daily life and health of residents in the basin region.

Village Background

Xiqiao Village is in the northwest area of Jiaxing City in Zhejiang Province, approximately 15 kilometres from the downtown area, 100 kilometres from Shanghai, 80 kilometres from Hangzhou and 70 kilometres from Suzhou. The village covers an area of 3.5 kilometres from north to south and 1.8 kilometres from east to west, totalling 4.37 square kilometres. Prior to May 2001, Xiqiao Village was under the jurisdiction of Rudong Township and was the seat of the Rudong Township Government. In May 2001, Rudong Township was merged with and absorbed by Xingcheng Town. Xiqiao, which originally had fourteen resident groups, merged with a nearby village. The newly formed Xiqiao Village has a total of 26 groups. As the seat of the Rudong Township Government, Xiqiao had already developed into a small market town. After the original government department disbanded in 2001, Xiqiao declined somewhat, but it still retained some of the characteristics of a market town, including a health station, post office, and other important institutions that were established during the time when the Township Government was located there. During this period, Xiqiao was comprised of three main streets. The village groups closest to these streets were Groups 4, 5, 13, and 14. A small river passes through the village from the north to the south (see Figure 5.1).

According to 2008 statistics, at the end of that year, Xiqiao had a total of 26 villager groups, 650 households, and 2,639 people. In fact, the actual population was larger than this because the village was also home to a large migrant population. Despite living in Xiqiao all year round, these people did not have household registration (*hukou*) there. From the 1990s, the migrant population of Xiqiao Village increased. From 2000, the migrant population was roughly equal to the local population and exceeded it in some years.[5] As a market town, Xiqiao was more economically developed than the neighbouring villages. As of October 2007, according to the Village Committee, there were 91 types of industrial enterprise (see Table 5.1), mostly located in groups close to the town. They included 81 textile factories operated by local villagers. It was these relatively developed rural industries that attracted the migrants, most of

5 The estimate of the migrant population in Xiqiao Village was made with the assistance of Mr. Xia, an accountant from the Xiqiao Village Committee, based on the Committee's statistical data and rental records for the migrant population. Although it is difficult to determine the exact number of migrants due to their high mobility, the accountant believes this is a reasonable estimate.

Figure 5.1 A schematic view of Xiqiao Village

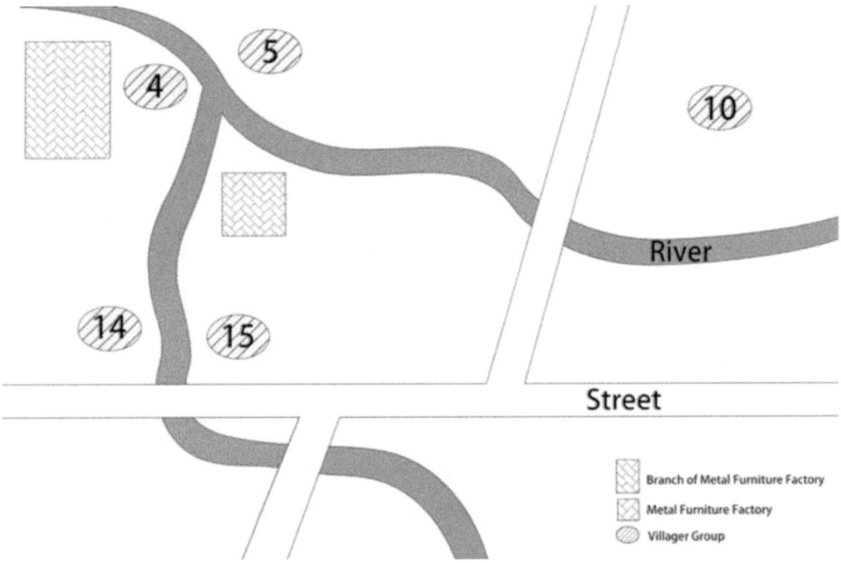

whom rented houses from villagers in the same groups around Xiqiao Market Town. To meet the demand generated by the migrants, villagers opened a variety of shops, and two relatively large supermarkets. The combination of industrial development and the influx of migrant workers meant that the income of Xiqiao villagers was quite high. Data shows that in the district where Xiqiao is located, rural net per capita income reached 12,500 yuan in 2009, and in the town, it was 12,000 yuan. Overall, Xiqiao is typical of the level of economic development of villages in the Lake Tai Basin region at the time in terms of per capita income and the living standards of local farmers.

Table 5.1 Industrial enterprises in Xiqiao Village

Type of Enterprise	*Number*	*Main Products*
Metal Furniture Factory	1	Outdoor leisure furniture (metal)
Water Jet Textile Mill	81	Grey cloth
Chemical Plant	4	Textile coating finishing agent
Clothing Factory	1	Wool sweaters
Tourism Products Factory	3	Tents and other travel products
Cooking Oil Plant	1	Cooking oil
Total	91	

Statistics source: Zhejiang County Second Economic Census (October 2007)

The development of rural industries caused severe environmental pollution in Xiaqiao, and especially water pollution. From the 1980s, the rivers in the village became increasingly polluted by waste water emitted by rural enterprises. Some of the smaller rivers deteriorated and became putrid, fetid ditches, posing a health risk to the villagers. Shallow wells used by village households also became polluted and were gradually abandoned. With the constantly increasing migrant population, the amount of household waste in the village was also mounting daily, and much of it ended up in the river. Over the previous decade, villagers had begun noticing an increasing number of people in the village dying of cancer, and some residents even began using the term 'cancer village.'

Dingbang is a natural village which is part of Xiqiao administrative village. Dingbang is located closest to the metal furniture factory, the largest polluting enterprise in Xiqiao, and it experienced severe pollution. Dingbang residents believed that the number of cancer cases and deaths in the village was higher than other villager groups, and for this reason the locals referred to it as a 'cancer village.' Dingbang consists of Groups 4 and 5. As shown in Figure 5.1, these two Groups are close to each other and separated only by a small river with bridges connecting the two sides. The population of the two Groups is also similar. According to statistics from 2005, at the time, Dingbang had 51 households and a registered population of 192 people, of which Group 4 had 26 households with 98 people and Group 5 had 25 households with 94 people.

Our research team conducted a series of studies in the area starting in 2005, including multiple field investigations of water pollution and disease in the village. This research delineates the history of water pollution in Xiqiao and summarizes descriptive statistics on the incidence and mortality rates for cancer in Dingbang. It also analyses the awareness of villagers towards the relationship between pollution and cancer, and their responses.

The Generation of Pollution and the High Incidence of Cancer

The History of Water Pollution

The Lake Tai Basin has historically been a region with a highly developed rural handicraft industry, with the silk industry having an especially long history. After the Tang Dynasty, the silk industry developed rapidly in the middle and downstream regions of the Yangtze River. Later, during the Song Dynasty, the Jiangnan region almost completely replaced the middle and downstream areas of the Yellow River as the centre of silk weaving in China. By the Ming and Qing Dynasties, almost every household in the Lake Tai Basin region planted mulberry trees and kept silkworms, and a group of silk cities appeared. From the nineteenth century, with the rise of national industries, modern silk weaving factories run by private entrepreneurs appeared in the Lake Tai region, but they were mostly located in large cities. Rural areas continued to focus primarily on traditional handicraft industries. Before modern factories appeared in the Lake Tai countryside, human interactions with nature could be described as reflecting 'harmony between people and water.' Traditional household silk weaving industries did not result in harsh water pollution, and the polders of the Lake Tai basin maintained a balanced ecosystem.

In 1958, Rudong Township began to establish People's Communes, and production teams, large and small, were called on to set up enterprises and start developing rural industry. Rudong Township established first silk mills and then brickyards. After the Household Responsibility System was established in 1983, the Township Government built glass factories and chemical plants as well. With this growth in industry, waste water produced by the factories ran into the river and polluted it. However, the production team enterprises established during the era of the People's Communes and the collective enterprises set up later after land was allocated to households caused only limited pollution. Although it did affect the water quality of nearby rivers, the impact was not severe, and some villagers still drank the river water directly. It was not until after the mid-1980s that villagers began drilling wells to obtain drinking water.

After the mid-1990s, waves of privatization swept over the Lake Tai Basin, first with the introduction of shareholding reforms in collective enterprises and then with total privatization. In the space of a few years, Rudong Township's communal enterprises changed hands many times and were quickly privatized, and as collective enterprises closed, private firms developed rapidly. After the privatization of Rudong Township's enterprises in 1997, Xiqiao also saw a period of rapid development and the fast growth of private enterprises.

This had two immediate consequences. The income of villagers increased noticeably, and water pollution worsened, resulting in general environmental deterioration. As in most areas of southern Jiangsu, the industries of Xiqiao were scattered, and regardless of the size or type of industry, enterprises were interspersed among residential homes. This made it difficult for the government to regulate emissions. During the early stages of rural industrial development, the local government did not put pressure on enterprises to regulate pollution, with the result that by the beginning of the twenty-first century, although industrial development had entered a golden era, water pollution in Xiqiao was reaching its worst levels.

Two Types of Heavily Polluting Enterprises: metal furniture and textiles

Of the 91 enterprises in Xiqiao, the metal furniture factory was the largest. It had the highest production value, and also the worst pollution. The owner of the metal furniture factory was initially a villager from one of Xiqiao's neighbouring villages. At the time of our research, the factory had four divisions. The main factory was situated in Xiqiao itself, and the other three divisions were located in the surrounding area (see Photographs 13 and 14). The factory mainly produced metal recreational furniture, using aluminium and iron as the primary raw materials. Its products included garden tables, chairs, furnaces, swing seats and sunlamps. Its products were mainly designed and produced for the European and American markets.

When the factory was first established, to reduce costs, scrap aluminium and iron were purchased for use as raw materials. After the enterprise expanded, scrap metal was no longer sufficient to meet production demands, so it bought in relatively good quality ready-made aluminium and iron blocks at higher prices. Originally, after the scrap aluminium and iron were brought to the factory, they were first put into a cleaning pool filled with sulphuric acid to wash away the rust and remove surface impurities; then the clean metal was put into a high temperature furnace to be melted (ready-made aluminium and iron blocks do not need to be cleaned, and go directly to this stage). The liquid metal from the high temperature furnace was then poured into various moulds and processed into prototypes of the target products. After these cooled, the surface was sanded, polished, and then spray painted. In the final step, decorative features were added. Figure 5.2 shows the production process.

Pollution from the metal furniture factory was generated primarily at three stages of the production process: first, in the cleaning pool where rust was removed; second, in the high temperature furnace which melted the metal; and third, during electroplating, spray painting, and decorating. The sulphuric acid used in the cleaning pool to remove rust interacted with oxides on the surface of the scrap metal, removing these and other impurities. In addition to sulphuric acid, diluted hydrochloric acid was also sometimes used for industrial rust removal. But regardless of which acid was used, the chemical reaction with oxides on the surface of metal produced either sulphates or hydrochlorides which are soluble in water and these chemicals entered the cleaning pool. The main workshop of the metal furniture factory had two cleaning pools, one for aluminium and one for iron. At a certain point, the sulphuric acid or diluted hydrochloric acid in the pools would be used up and only waste water and residue remained. These were then emitted into the

Figure 5.2 Production process at the metal furniture factory

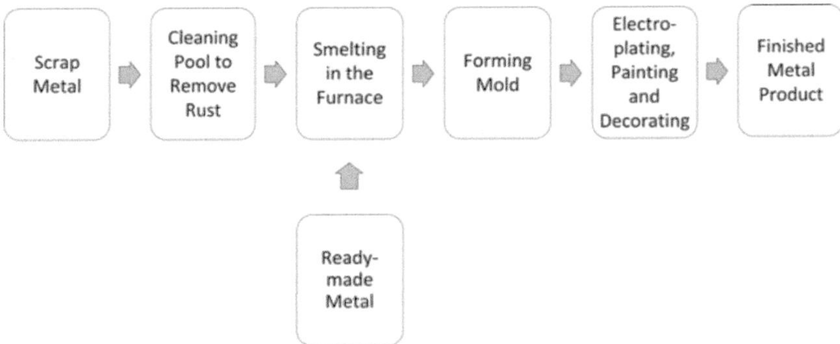

river through pipes, causing pollution. Chemical substances and residue in the waste water turned the river a bright colour, and a lot of toxic substances were deposited on the river bed. The early years of the metal furniture factory's operation, when a lot of scrap metal was used to lower production costs, was the period when the factory caused the most serious pollution.

Another stage in the production process which caused pollution was the high temperature smelting. Scrap metal which had undergone rust removal entered the high temperature furnace to be melted together with metal purchased from outside. The solid metals were liquefied and fused, ready for the next stage of forging and moulding. Because the melting points of iron and aluminium are both very high (the melting point of iron is around 1,500°C, and aluminium around 700°C), the furnace had to be extremely hot. As they melted, iron and aluminium produced iron oxide and aluminium oxide. Iron oxide is a black solid and does not dissolve in water, and aluminium oxide is a white powdery substance. A large proportion of these two substances was expelled through the chimney of the furnace in the form of smoke dust when the furnace was in operation. The chimney was not very high (roughly 20 metres) and near the river. The smoke dust was very dense and black, and an unpleasant sight, and because it provoked the anger of the villagers, the furniture factory usually only operated the furnace at night. The villagers most affected by pollution were those in Groups 4 and 5 of Xiqiao, which are the closest to the main factory, just across the small river, which is less than 20 metres wide.

The electroplating process also caused heavy metal pollution. Electroplating is a process that uses electrolysis to coat the surface of one metal with another layer of metal or a metal alloy. At the time, electroplating of metal furniture in China commonly used the heavy metals zinc and chromium, which can damage health.

The second largest source of pollution in Xiqiao was the huge number of household textile mills. At the end of 2008, the village had a total of 650 households, distributed among 26 Groups, and there were 81 water jet textile mills (see Table 5.1), or an average of one mill for every eight households. These mills were mostly distributed on the two sides of the Xiqiao Village road, which is also where Groups 13 and 14 were located. Only a few mills were in the other 24 Groups. Jet looms use water jets to draw wefts through the shed of looms that do not have shuttles. This means that they require a tremendous amount of water. Each loom used about three to four tons of water a day. When there was an adequate supply of raw materials and consistent demand, the mills in Xiqiao operated around the clock with no rest period. Calculating that each household had an average of ten looms, and each one used four tons of water per day, then Xiqiao was using a total of 3,240 tons of water per day. From a practical point of view, even though Xiqiao is rich in water resources, the surface water was not sufficient to support this number of water jet mills. The mills that were built earlier were usually closer to the river, and typically used river water. But those which were built later usually dug deep wells inside the mill area and extracted ground water. Regardless of whether they used surface water or ground water, all the mills had built waste water channels that ran into small rivers. Usually mills that were built earlier used open-air pipes to emit waste water (see Photograph 15), while those that were built later used underground pipes. Of the water being used in loom operations, the fabric absorbed 8-10%, 3-5% evaporated into the air, and the remaining 85-87% was discharged through the gutters in the workshop into the waste pipes. If 3,240 tons of water were being used in normal daily operations, subtracting the amount taken by the fabric and evaporation, the final amount of waste water produced would be 2,800 tons. Of this, most was discharged into nearby rivers, and a small amount seeped into the ground water. What chemical substances did the waste water contain? In the textile industry, there was one important step before the water jet looms; warp sizing, or sizing for short. The warp sizing process is called the 'tiger's mouth' by the textile workers because this term vividly captures its importance in the industry. This step is so important because when the warp yarn[6] undergoes the repeated impact of the loom, the surface can feather, and bobbles may form. The yarn may even break. To ensure that the warp yarn will function normally on the loom, sizing paste must be applied. At the time, the main ingredients of the paste used

6 The warp yarn is the yarn that is held in place on the loom, while the weft yarn is woven above and below it.

in warp sizing in China included modified starch, polyvinyl alcohol paste, and acrylic paste.[7] Modified starch is a derivative product prepared using physical, chemical, and enzymatic methods to change and enhance the natural properties of starch and expand its scope of application.[8] Modified starch is still a natural fibre made from plants, and so its environmental impact is much smaller than that of chemical pulp. Polyvinyl alcohol on the other hand is a kind of synthetic polymer. A white powder, it is soluble in water, but only at temperatures of about 65-75°C. Polyvinyl alcohol size is not easily biodegradable and causes serious environmental pollution. Acrylic size is a colourless liquid. It does not cause pollution, but it is highly absorbent and prone to becoming sticky again in the weaving process. For this reason, it can only be used as a supplemental material in combination with other types of size. At the time, in China, starch accounted for around 70% of size used in textile production, polyvinyl alcohol for around 20%, and acrylics and other types of size for around 10%.[9]

Jet looms use water jets to insert weft yarn, and so part of the warp yarn gets soaked with water. Even though the warp yarn has gone through sizing, and the size itself has a certain level of water resistance, a certain amount of size will wash off the warp yarn, enter the gutters in the workshops and then the waste pipes. In Xiqiao, the waste water discharged from all the water jet looms came out white, forming a thick layer in the channels it flowed through. This white sediment was mostly sediment from chemical pulp.

Another substance was in the waste water discharged from water jet textile mills – engine oil. Oil was applied to the looms to ensure smooth operation and also to reduce wear and tear. A small amount of this oil also dripped into the gutters when the machines were running and was discharged together with the waste water. Water jet looms typically operated 24 hours non-stop, using a large amount of oil, and so quite a lot of oil was discharged in the waste water. In fact, the waste oil discharged by water jet textile mills had already spawned a related subsidiary industry. Xiqiao villager Mr. Xia described the situation.

> Every loom needs oil. For example, after one day of the waste water from 40 mills has been processed there's a dozen kilograms of oil. If it's not processed, the oil enters the river together with the water, and when it

7 Zhou, 'Modern Textile Warp Sizing'.

8 Production Process of Modified Starch.

9 Wu Hailiang et al., 'Development of Textile Sizing Agents and Environmental Protection', pp. 12-14.

> enters the paddy fields, the rice saplings will die. We have a few Anhui people here who specialize in going round the mills to scoop the oil. Each person can scoop a few dozen kilograms of oil each day, and the best can get over 50 kilograms. Between them, a few people can scoop 200 to 300 kilograms of oil per day. They can sell the waste oil to places that specialize in buying it for more than one yuan per half kilo. Usually they go every two or three days to a mill to scoop oil. Some mills have started to scoop the oil themselves after seeing that they can get money for it, but most owners don't care for small change, so they let others scoop it. Anhui people did this for a few years over here, and got rich, and now they drive cars to come scoop the oil.

Even though some of the waste oil was scooped away and sold, some of it still drifted with the waste water into the river. Because the density of oil is lower than the density of water, and it does not dissolve, these waste oils floated on the surface of the water, sparkling in the sunshine. (See Photograph 16).

Waste water discharged by the metal furniture factory and the numerous water jet textile mills quickly caused the rivers around Xiqiao to lose their original functions. The river that passes through Xiqiao Village was the most severely polluted because more than 80% of the all factories in the village were clustered on its banks. This small river lost almost all its natural functions, becoming a waste pond for the factories and a garbage pond for the residents. Every day a large amount of waste water was discharged into this small river and many residents also threw their household garbage in there. The water in the river was partly white from the textile mill emissions, but the water discharged by the metal furniture factory was sometimes black and sometimes other colours. Local villagers often described their frustration with water pollution by saying, 'If anyone can find river snails in the river, I dare to eat them raw.' River snails have a strong will to live and are very disease resistant. They reproduce rapidly and can survive in water of quite poor quality. If even the river snails are dead, it means the water pollution is incredibly severe. Mrs Ma, a 50 year-old woman who lived by the river, was full of sarcasm and anger when discussing the water pollution.

> See how funny the water in our river is? it has turned so colourful. You can't see this anywhere else! In the past, the water we had was good, the water we drank all came from the river. In summer when you came home from work, you'd go to the river to wash yourself. It was very refreshing. Ever since the metal furniture factory was built it was no good anymore [pointing across the river], the river began to turn black. And after that,

> those textile mills were also built, and the river turned white. The water they discharge has engine oil in it and now the river water has colours. Since tap water was installed in 1997, we haven't used the water from the river. The water is no good – it's not fresh. The ducks we kept all died. They couldn't go into the water. Ugh, you can't even wash out a mop in the river, it's all stinking. We're not happy to be by to the riverside [...], but what can we do. Our house is right by the river.

The Cancer Situation

The people most affected by pollution from the metal furniture factory were the two groups in Dingbang Village. These two groups also had a higher incidence of cancer and a higher mortality rate than other groups, and were called a 'cancer village' by villagers from other groups. Through household surveys and cross-checking, our team collected information about the incidence and mortality from cancer among Dingbang villagers. Table 5.2 reports the data for recent years and Table 5.3 shows the list of villagers who contracted cancer but did not die. It is important to note that when investigating cancer, the team did not request patients' medical history to verify whether their account of their situation was true. However, we used various methods to verify these claims and we believe that the data is authentic. In the countryside, to have cancer is a tragic event and a tremendous blow to the patients themselves and their families. To have family members contract cancer is considered shameful, and so no villager would falsely claim to have cancer themselves or make similar claims about their family members or other villagers.

As can be seen from Tables 5.2 and 5.3, most Dingbang villagers died of cancer between 2002 and 2003, accounting for nearly half of all deaths from 2000 to 2009. In terms of the type of cancer, villagers mostly contracted cancers of the digestive system, including oesophageal and stomach cancer. Research has found that there is an association between environmental factors and the high incidence of cancer in China in recent years, and that the relationship between digestive system cancers and environmental pollution is especially close. Carcinogens in the environment enter the human body through drinking water and food, causing the development of cancers, especially oesophageal, stomach cancer, liver and other digestive system cancers.[10]

10 Liu et al., 'Etiology of Guangdong Villages with High Incidence of Digestive System Malignant Tumors', pp. 1139-1141.

Table 5.2 Cancer mortality statistics for Dingbang Village

Serial Number	*Patient Family Name*	*Gender*	*Year of Death*	*Age of Death*	*Type of Cancer*
1	Lu	Female	2000	57	Stomach cancer
2	Chen	Male	2001	73	Colorectal cancer
3	Chen	Female	2002	73	Breast cancer
4	Shen	Male	2003	54	Oesophageal Cancer
5	Shen's Wife	Female	2003	40	Lung cancer
6	Chen	Male	2003	80	Oesophageal cancer
7	Jin	Male	2006	55	Oesophageal cancer
8	Chen	Male	2006	60	Oesophageal cancer
9	Chen	Male	2007	57	Stomach cancer

Table 5.3 Dingbang Village cancer patients who are still alive

Number	*Patient family name*	*Gender*	*Onset year*	*Onset age*	*Cancer type*	*Village Group*
1	Chen	Male	2002	51	Stomach cancer	Group 4
2	Chen	Female	2008	62	Nasopharyngeal cancer	Group 5

Dingbang Village includes two Groups. For many years, its population was stable at between 195 and 200. It can be seen from Table 5.2 that in the seven years between 2000 and 2007, nine people in Dingbang died of cancer. Given that the population in 2005 population was 192, the average cancer mortality rate between 2000 and 2007 was 586/100,000. Table 5.4 lists the national rural cancer mortality rate for China in the same period. We can see that the annual average cancer mortality rate of Dingbang Village was approximately four times that of the 2007 national rural cancer mortality rate, which was, at its peak in 2007, 144.2/100,000.

Epidemiological studies often use case-controls to examine the association between certain factors and disease, and the strength of the association.[11] Using this principle, we chose a relatively less polluted group in Xiqiao and compared its cancer mortality rate with that in the two groups from Dingbang.[12] Based on its distance from the pollution source, we chose Group 10 as the control group. As we can see

11 Zhang, 'Practical Modern Epidemiology Methods', pp. 315-317.

12 Subjects of case-control studies in epidemiology and its research methods have strict requirements, including sample size amongst others. Here I only borrow this principle to analyze

Table 5.4 China annual rural cancer mortality rate (unit: 1/100,000)

Year 2000	*Year 2001*	*Year 2002*	*Year 2003*	*Year 2005*	*Year 2006*	*Year 2007*
112.57	105.36	83.34	84.3	107.0	130.2	144.2

Data source: People's Republic of China Ministry of Health website www.moh.gov.cn

from Figure 5.1, Group 10 is located further from the town, and the river that runs through the village from north to south also does not reach Group 10, so residents were less affected by pollution from the river. Another important factor was that Group 10 was a long way from the metal furniture factory, and mostly not affected by its pollution. I also conducted household investigations for Group 10 and compared the data many times. The resulting data for cancer mortality from 2000 to 2007 are shown in see Table 5.5.

Table 5.5 Cancer mortality in Group 10, Xiqiao Village

Serial Number	*Patient family name*	*Gender*	*Year of death*	*Age of death*	*Type of cancer*
1	Chen	Male	2004	67	Stomach cancer

According to my analysis of the population register of Group 10, the population remained stable at around 100 people for many years. Assuming that the average population of Group 10 between 2000 and 2007 was 100, then the cancer mortality rate was 142.86/100000. This number is slightly lower than the 2007 national rural average, and at a normal level. Comparing the cancer mortality rate of the control Group 10 with the two groups in Dingbang shows that the cancer mortality rate in Dingbang was 2.72 times that of the control group. Group 10 and Dingbang are both in Xiqiao, and so there is no large difference in villagers' living habits, or other geographical factors that would influence their health. Does the large difference in cancer mortality rates prove that there was a close association between industrial pollution and the high cancer rate in Dingbang? Inevitably, the difficulty in determining the population base and identifying pollution sources accurately means that our analysis can only be exploratory.

the data, and do not claim to have determined the relationship between cancer and pollution from an epidemiological perspective.

Villager Awareness of the Relationship between Pollution and Disease

The Path to Awareness

Awareness of environmental health risks has three aspects: awareness of pollution, awareness of disease, and awareness of the relationship between the two. Because cancer has an extremely high mortality rate and is regarded as an incurable disease, villagers are very sensitive about it. But Dingbang villagers' awareness of cancer was limited to its high mortality rate and the fact that it is difficult to cure, and not much else. When asked 'how much of a connection is there between pollution and cancer?' the response of most villagers was 'we cannot tell.' Nonetheless, some cancer patients and their family members were convinced that the high incidence of cancer among villagers was caused by pollution from the metal furniture factory, and that it had a close connection with the poor environmental quality in the village.

Knowledge about pollution, about cancer, and about the relationship between the two, is all very specialized, and Dingbang villagers' path to awareness was predominantly through their daily life experiences. This included their sensory and physical contact with polluting substances in daily production and life activities, which led them to directly experience the potential harm that pollution can cause the human body. For example, after the river water was polluted, when some farmers washed their hands and feet in the river after work, they immediately developed a red rash that was incredibly itchy and uncomfortable. This had never happened before, and so the villagers deduced that the river water might contain harmful chemical substances. Additionally, they discovered that after the pollution in the village worsened, the number of fish, shrimp and other aquatic creatures in nearby rivers and lakes declined. By the time of our research they had almost completely disappeared from the smaller rivers around the village. Aquatic animals are an important living indicator of changes in water quality and through the effects of pollution on them, the villagers also perceived the threat of pollution to human health. Again, the chimneys of the metal furniture factory often released a dense black smoke and many villagers in Dingbang dared not open their windows at night. When they woke up in the morning, they could detect a distinctly strange smell, and the leaves of the vegetables growing in patches beside the river were also often covered with black particles.

Through such daily life experiences, villagers became more aware of the health risks caused by pollution and came to believe that the high incidence of cancer in the village in recent years was a direct result of environmental pollution. However, these perceptions based on everyday life experienced still belonged to realm of feelings, and often did not receive recognition from professional institutions due to the lack of scientific evidence.

Monitoring of pollutants is usually the responsibility of the Environmental Protection Bureau (EPB). The local EPB did once pay a visit to Dingbang and the metal furniture factory, but whether they conducted tests for pollutants, and what the results were, the villagers never got to know. Monitoring the incidence of major diseases in the population is the responsibility of the Centre for Disease Control (CDC), which did not conduct any investigations into the incidence of cancer in Dingbang. In fact, the 'high incidence of cancer' that Dingbang villagers believed to exist was actually very difficult to determine using epidemiological methods. This was partly because Dingbang's population was too small for epidemiological research, but if the population base was set to the entire Xiqiao Village, then the result would be diluted. Cancer in Dingbang could also have been caused by other factors.

In 2010 we visited the local EPB and the CDC. Employees of the CDC told us that the city had already built a surveillance network and reporting system for major diseases, but that from an overall point of view, there was no high incidence of cancer in the city as a whole. However, the staff did also believe that regulating the environment was the premise for improving public health, and that this could only be done by improving the ecological quality of the environment.

Awareness of pollution sources and environmental media

Environmental pollution in Xiqiao included industrial pollution, agricultural pollution, and household pollution. All of these caused environmental and ecological degradation, but the one that affected the villagers' health most severely and directly was industrial pollution, because these discharges contained hazardous chemical substances. When I first began research in Xiqiao, I was not aware of the high incidence of cancer in Dingbang. On 3 May 2008, when I was chatting with the owner of a grocery store by the river, she said that her family had moved to that street from Dingbang, where a lot of people were dying of cancer in recent years. She said the locals all referred to Dingbang as a 'cancer village,' and that the residents on the street were unwilling to buy rice and vegetables from there. This was when I noticed Dingbang and started researching the high incidence of cancer among its villagers.

Dingbang villagers commonly believed that the pollution discharged by the metal furniture factory was the main cause of the high incidence of cancer. They gave three reasons for this:

First, the waste water and gas discharged from the factory could contain toxic substances. The villagers believed that even though the water discharged from textile mills was white, its main content was still textile size, and this usually would not contain toxic chemicals. Because many village households had textile mills, they knew a certain amount about sizing agents. But in fact, they were not completely correct about the waste water. As discussed above, three main types of textile sizing agents were in use in the Chinese textile industry at the time: modified starch, polyvinyl alcohol, and acrylic paste. Modified starch was made mostly from plant-derived natural fibres and not toxic. Polyvinyl alcohol does not easily dissolve naturally and has a certain level of toxicity; and acrylic paste does not pollute the environment. The size used in Xiqiao contained a certain percentage of polyvinyl alcohol, and so the waste water discharged from the mills still contained toxic substances. Waste water discharged from the metal furniture factory was often black or other colours, and the villagers speculated that it was the liquid waste that remained after washing scrap metal with diluted sulphuric acid and so it contained residues of these metals. Other than that, the smoke being expelled from the factory chimney had a pungent smell. A few villagers from Dingbang used to work at the factory and learned about the exact production methods. Furthermore, after returning home after a night shift one of the villagers suddenly died of a stroke, and another also suffered a sudden stroke after working for a few years at the factory. Although he did not die, he was disabled, with severe hemiplegia. Two villagers dying or becoming disabled after working at the metal furniture factory served as a warning to other villagers, and very few of people from Group 2 in Dingbang worked there.

Second, Dingbang was most severely affected by pollution from the metal furniture factory. There were divisions of the factory on both the east and west sides of the village, and so the hazardous waste water and gas reached it first. The fields and vegetable patches of Dingbang were mostly situated near the village, and the irrigation water came directly from the river, so the poisonous waste water discharged by the metal furniture factory affected the rice and vegetables planted by Dingbang villagers most. The fields and vegetable patches of other groups were slightly further away from the furniture factory so they could use water from other rivers for irrigation, and pollutants discharged from the metal furniture factory might be more diluted by the time they reached them. Dingbang was situated in the fork between the main metal furniture factory and its branch factory, and so the smoke and dust emitted from the factory also affected Dingbang the most. In the years when the smoke and dust were especially severe, Dingbang villagers did not even dare to open the windows when they were sleeping at night.

Third, the timing of the onset of cancer and death among Dingbang villagers was basically consistent with the development of the metal furniture factory. The main plant was built in 1998, and the branch factory near Dingbang was built in 2002. The onset of cancer and death among villagers was mainly concentrated in the period after 2000 (see Table 5.2). The villagers believed that after 2000, output from the factory was high and its scale increased rapidly; furthermore, before 2005, waste water was discharged directly into the river without treatment. After 2005, due to the strong opposition of Dingbang villagers, the factory finally paid to have waste treatment equipment installed, but secret dumping of waste still occurred from time to time. The villagers generally recalled that before 2000, it was rare for Dingbang villagers to have cancer. In May of 2003, two people in the village fell ill in quick succession and were diagnosed with cancer after check-ups in Hangzhou. In the years that followed, there were successive cases of people contracting cancer. This caused panic among the Dingbang villagers.

Because the river runs directly through Xiqiao, in principle, all villagers living by the riverside should have been victims of pollution. But toxic substances are most concentrated right after being discharged. As pollutants are diluted by river water or sink to form sediment, theoretically, residents further away from the metal furniture factory should be less affected by the pollution. Looking at the distribution of the village and the distribution of the metal furniture factory (see Figure 5.1), the two villager groups of Dingbang were exposed to the greatest harm from pollution. How did these toxic substances enter the body and cause harm to health? Dingbang villagers believed that they did so through three environmental media – water, food, and air.

Water is the substance with which villagers have most contact in their daily lives. Villagers believed that they were exposed to poisonous substances in the water through drinking it and washing food. Rapid industrialization came to Xiqiao in the mid to late 1990s, and this was when the general deterioration of the water started. Before this, drinking water and water for domestic purposes was all taken directly from the rivers. By the 1980s, the water quality in nearby rivers had started to deteriorate, and it was no longer suitable for drinking. Villagers began to dig wells in their own homes to get water. By the late 1990s, the rivers near Xiqiao were already heavily polluted, and the well water in the villagers' homes was also affected by pollution. In 1997, the village collectively funded a 200 meter deep well to be dug near the Xiqiao Village road, solving the drinking water problem for nearby villages.

But solving the drinking water problem did not entirely prevent exposure to pollutants through water. On the one hand, villagers who used tap water had to pay, and villagers with relatively low incomes still used the polluted river water and well water to wash food, including rice and vegetables, to save money. In addition, some villagers ate polluted fish and shrimps and other aquatic products. And even if water for daily use came from tap water, river water was still needed to irrigate the fields and pollutants got into the crops in the process. Toxic substances can remain in the soil for a long time, eventually being absorbed by food crops. While the majority of villagers realized that the food and vegetables they grew were polluted, they had to keep eating them. But when they took surplus grain and vegetables to market, if the buyer knew the seller was from a 'cancer village', they would refuse to buy.

Air was also an important environmental medium for spreading pollutants. The metal furniture factory in Xiqiao built a high temperature furnace by the river. The chimney was very low and in the first few years after the factory was constructed, it discharged thick black smoke even in daytime, Dingbang villagers across the river were seriously affected, and did not dare open their windows even during the day. Later, due to opposition from villagers, the factory changed to operating the furnace at night, usually after nine o'clock. But the black smoke not only directly polluted the nearby air, it also deposited black particles on the surface of rice and vegetables planted by farmers.

Villagers' Responses to Health Hazards

Resistance

After realizing the health hazards associated with industrial pollution, Dingbang villagers began to take action. Their main demand was that the factories should stop polluting and they expressed this in three ways: by negotiating with the enterprise; reporting the enterprise to the government; and involving the media.

In one week during May of 2005, two people from Dingbang Village were diagnosed with late-stage oesophageal cancer. Villagers began to realize the seriousness of the problem and organized to go to the metal furniture factory to negotiate with the director, Mr Xu. But when they arrived, they could not find him. Instead, other leaders of the factory received the villagers, and made all kinds of excuses to shirk responsibility. The villagers did not receive a satisfactory answer. Later, they turned to the retired former township Party Secretary, Mr. Chen. Chen was from Dingbang and was living in the village after retirement. When the villagers came to him, he wrote a document titled, 'Reporting a Situation, and Making a Suggestion', and entrusted it to someone to deliver to the director of the furniture factory. But he also did not receive a reply.

When direct negotiations with the factory did not produce results, the villagers began to report the problem to the Village Committee and the local government. They first reported the situation to the Party Secretary of the Village Committee, but his reply was: 'There's nothing we can do.' At the end of 2005, in the retired cadres' discussion meeting, former secretary Chen reported the situation of pollution and cancer in Xiqiao to the township Party Secretary and other cadres at the meeting said: 'This type of factory must be shut down.' But the Party Secretary did not take a position. Afterwards, Chen also represented the villagers in reporting the problem to the deputy Village Head in charge of industry, but he also did not take a position. When reporting to the township Party and government organizations both proved ineffective, some villagers used the 'mayor's hotline' to report the pollution situation in the village. The city instructed the district Environmental Protection Bureau to send people down to investigate. However, after the inspection, the EPB concluded that even though the metal furniture factory discharged pollution, the pollutants did not contain toxic substances, and had no direct connection with cancer in Xiqiao.

Xiqiao villagers also sought help from the media by calling a live chat show on the local TV station. A reporter from the show went to Xiqiao to conduct interviews, but afterwards villagers did not see the program on the air. They called the TV station again, and the reporter told them that the program had not been approved and could not be aired. From this process it is clear that the village gentry and village elite still play a role in rural collective affairs. The former secretary, Mr. Chen, acted as a township cadre from 1975 to 2001, and after retirement in 2001 he returned to live in his home village. As a retired cadre, he was very familiar with the current township leaders, and had channels to represent the villagers' concerns. His position as a member of the 'village gentry',[13] meant that

13 Wu and Fei et al., *Imperial Power and Authority of the Gentry.*

the old secretary had a certain prestige among the villagers, and they were willing to report problems to him. Throughout their resistance to pollution, the Dingbang villagers never resorted to extreme actions, and did not 'naoshi' (stir up trouble). The former Secretary believed that he played a role in this by advising villagers to report the problem through the proper channels, and not be confrontational. Dingbang villagers also believed that the people in the village wanted a quiet life and were unwilling to make a fuss. But at the same time, the former secretary's own intervention did not get enough attention among the current leadership to produce a solution.

Strategies of Non-Resistance: avoiding harm and seeking to escape

After active resistance failed to bring results, some villagers began to take other measures to avoid harm. Their main strategy was to try to avoid coming into contact with pollutants. Water is an important medium of exposure for toxic substances, and so they tried to avoid contact with polluted water as far as possible. As early as the late 1980s and early 1990s, after villagers discovered that the river water was getting dirty, they began digging wells in their own yards, and switched their drinking water from river water to underground water. By 1997, the rivers in the village had completely deteriorated, and the Village Committee raised funds to dig a 200-metre well, enabling the whole village to use tap water. However, because there was a fee for tap water, many villagers still used their own well water for daily washing purposes. After 2000, when villagers discovered more and more cancer patients, many also stopped using well water to wash vegetables and rice, and only used it to wash clothes. In fact, since the switch to tap water in 1997, most villagers had already cut the connection with the rivers outside their doors. Many villagers reported that the river water could not even be used to wash mops and had a very strong stench. However, agricultural irrigation continued to require a large amount of water, and the villagers had no choice but to use river water for crops and vegetables. Only a handful of villagers used well water instead of river water to irrigate the vegetables they grew for their own consumption.

Ever since the abnormally high incidence of cancer appeared among Dingbang villagers, locals started to be suspicious of the crops they produced. When they brought their homegrown rice and vegetables to local markets, many non-Dingbang people were unwilling to buy them. Some villagers also suspected that their homegrown rice was poisonous, and a small number of villagers who were in a better financial situation bought rice from elsewhere.

For the majority of villagers, however, selling their homegrown rice and then buying rice from elsewhere was not financially feasible, and so they still ate their own rice. In terms of the distribution of risk, Dingbang villagers were therefore the major victims of pollution and bore the biggest health risk. However, when contaminated rice enters the market and circulates beyond the area, the risk is actually spreading, becoming 'dispersed' across the entire society. This also conforms to Beck's concept of the allocation of risk in the 'risk society'.[14]

In addition to avoiding contact with pollution sources, some villagers began to try to escape it permanently by moving away from the village. In Dingbang Group 5, between six and eight of a total of 25 households had already bought houses (at commercial rates) in the town or urban districts. For these people, the decision to buy a home in town rather than building a new home back in the village was often mainly due to their financial situation or job requirements, but serious local environmental pollution was also an important factor that they took into consideration. A phenomenon related to the purchase of property outside the village, was that in the previous ten years, not a single household in Dingbang Group 5 built a new house in the village. The few villagers who did build new homes chose places far away from the village. When building a house, most villagers first considered affordability and other factors, but they commonly agreed that even if they were to build a new house, they would not choose to build it in the village, and they hoped that their new home could be far away from the polluting factory.

Mr Chen, a resident of Dingbang Group 4, had contracted stomach cancer eight years previously and recovered. He firmly believed that his cancer was related to pollution. His son worked in the city, and his granddaughter basically never came to the grandfather's home, but frequently went to visit her maternal grandmother, because the environment was better there. Escaping from the source of pollution was a choice made out of the villagers' frustration and demonstrates their concern about environmental health risks. Because this practice can also lead to the hollowing out of the village community and many other social problems, pollution is not only leading to disease and health damages, but in the long run it will also affect economic development and cause the living standards of residents to decline. Villagers tried to use daily life practices to prevent pollution from damaging their health. But if pollution is not effectively controlled for a long period of time it will naturally cause other social problems.

14 Beck, *Risk Society: Towards a New Modernity.*

Factors Shaping Resistance Strategies: beneficiaries and victims, face and favours

The concept of benefit and victim circles is a classic theory from Japanese environmental sociology, which originated from the empirical research of Japanese sociologists working on the problem of the public nuisance caused by the building of the high-speed rail system (Shinkansen). The benefit circle refers to the people who benefit from the public nuisance, and the victim circle refers to the people who are its victims. The benefit/victim circle approach requires identifying these two different types of people, and analysing the relationship between them to understand whether they are separate or overlapping.[15] We can use this approach to analyse the problem of industrial pollution in Xiqiao Village. Who benefited from the development of the enterprises that discharged waste and caused environmental pollution? Clearly, the main beneficiaries were the owners of the enterprise, and the people working there were next. In analysing the victim circle of industrial pollution in Xiqiao, we found that the major victims were the ordinary residents of the village, including both locals and migrant workers. Due to their high mobility, migrant workers were only temporary victims of pollution, but local residents were permanent victims. Since the owners of the Xiqiao textile mills were all locals, they were also victims of pollution, but the owner of the metal furniture factory and his family did not live in Xiqiao and so they were not exposed to pollution. In sum, most enterprise owners in Xiqiao were both beneficiaries and victims of industrial pollution, and only the owner of the metal furniture factory was solely a beneficiary.

Two points are worthy of note here. The first is that the beneficiaries of pollution ranged from the enterprise owner to factory employees, and then ordinary residents, and the degree of benefit gradually decreased. It was greatest for the enterprise owner. Employees working at the factory derived less benefit because they only received income, and ordinary residents derived the least benefit. In reverse, the victims of industrial pollution ranged from the enterprise owners to factory employees, and then ordinary residents, and the degree of victimization gradually increased. Because enterprise owners benefited most from the factory, they also had more options to effectively avoid the harm caused by pollution. Ordinary residents were exposed to the greatest harm and were least able to avoid it. The second point is that industrial development in Xiqiao and associated pollution caused social stratification among village residents. They could be divided into three

15 Bao, 'Sociological Theory of Environmental Issues – Research by Japanese Scholars', pp. 89-94.

levels according to their income. The enterprise owners who first became rich because of industrial development were at the top; skilled workers and ordinary residential households with more workers were next; and families with fewer workers or whose health had been damaged by pollution were last. A small number of families had fallen into poverty because the main worker in the family died of illness or lost the ability to work, and these families were the most vulnerable in Xiqiao. They could not enjoy any additional income from industrial development, and they also suffered its negative impacts. This phenomenon of 'poverty within prosperity' deserves our concern.

In the case of Xiqiao, the overlap between benefit and victim circles influenced pollution victims' selection of strategies for resistance and their intensity. Villagers, especially those with family members working in the factories, were mostly unwilling to initiate environmental resistance when they experienced harm from pollution, or at least not direct resistance. In Xiqiao, salaries at the local factory were lower than those in factories across the river, so villagers who stayed in the village to work usually did so because they were unable to leave their homes for some reason, or they had limited skills and could not find employment elsewhere. For these villagers, if they were to lose their jobs because of environmental resistance, it would be a disaster for their families.

In Xiqiao, the tight overlap between the benefit and victim circles complicated the villagers' environmental resistance and presented a particular set of difficulties. In addition to this, in a 'familiar society' where everyone knows everyone, personal relationships and 'face' were also important considerations. Even though Xiqiao is quite modernized, and huge changes have already taken place in its social structure, overall, the characteristics of a 'familiar society' were still apparent.[16] Complex kinship and spatial relationships exist between the villagers. Even though ordinary residents were the greatest victims of pollution, they knew the enterprise owners, and if conflict and disagreements arose over pollution, it would damage their social relationships. Because of these personal considerations, ordinary villagers were usually unwilling to argue with their neighbours about pollution. Also, the environment is a public good, and for villagers to start disagreements with people in the community over the 'public interest' is not in accordance with the ethics and morals of traditional society. In this context, personal bonds and face restrained victimized villagers from engaging in fiercer environmental resistance.

Because the owners of the textile mills were all locals, to avoid speaking ill of them, villagers were typically unwilling to bring up the subject of textile mill pollution. Since the owner of the Hero Metal Furniture Factory was

16 Fei, *Rural China Birth Control Policy*, pp. 6-11.

from a neighbouring village, even though the Xiqiao villagers knew him, he was nonetheless not a local, and so they typically attributed most of the responsibility for pollution to the metal furniture factory. In Group 5, there was a family with the surname Chen whose house was directly connected with the neighbouring textile mill. When the looms were in operation, the vibration from the machines caused the Chen house to vibrate along with them and the noise was very loud. Yet when we chatted with Mr. Chen, he constantly talked about pollution from Hero Metal Furniture Factory. When we asked whether production from the neighbouring textile mill disturbed his daily life, Mr. Chen said only, ‘it’s alright when you get used to it’. But while we were interviewing other villages in their homes, his daughter-in-law came over specifically to tell us about the disturbance caused to their family by the neighbouring textile mill:

> These few years, we can’t eat well, we can’t sleep well, and even our bed is moving. That house of ours is cracking apart. This isn’t like a home. Home is a peaceful harbour. What kind of situation is this? It’s like this all year around. Our blood pressure has gone up, and our nerves are frayed.

When the daughter-in-law of the Chen household was talking to us, the father of the owner of the textile mill next door to them also came by, and she immediately fell silent. Afterwards she told us they are all next-door neighbours, and it is not good to complain face-to-face, plus the father of the mill owner was the leader of their Group. These frequent encounters in a familiar society where everyone knew everyone made the villagers turn a blind eye towards pollution from the textile mills. But they also faced obstacles in attacking pollution from the furniture factory. Because it was relatively large, the enterprise was one of the largest taxpayers in the region, and the owner had a relatively high social position and was influential. Ordinary villagers even did not have the chance to meet him personally.

The overlap between the benefit and victim circles, along with concerns over personal bonds and face among acquaintances, meant that people in Xiqiao very rarely initiated direct and intense confrontation with enterprise owners over pollution. Victimized villagers preferred to choose a more roundabout form of resistance, such as reporting to government departments. However, after repeated reporting, villagers were becoming disappointed at the attitude and action of the government. They believed they basically did not care about environmental pollution: and even if they went to investigate, it would not amount to anything in the end.

Works Cited

Bao, Zhiming. 'Sociological Theory of Environmental Issues – Research by Japanese Scholars'. *Xuehai* 2 (2009): 89-94.

Beck, Ulrich. *Risk Society: Towards a New Modernity*. London: Sage Publications, 1992.

Fei, Xiaotong. *Rural China Birth Control Policy*. Beijing: Beijing University Press, 1985.

Gao, Junfeng, Mao Xinwei, Yan Zhijun, and Zhong Huaping. 'Lake Tai Basin Economic Development and the Demand for River Resources'. *Science & Technology Review* 2002 (issue 4): 62-64.

Liu, Yisheng, Gao Yi, Wang Kangwei, Mai Xiaohan, and Chen Guangdao. 'Etiology of Guangdong Villages with High Incidence of Digestive System Malignant Tumors'. *China Tropical Medicine* 2005 (issue 5): 1139-1141.

Wu, Hailiang, Yang Mingkel, Shen Yanqin, and Cheng Xuezhong. 'Development of Textile Sizing Agents and Environmental Protection', *Cotton Textile Technology* 2008 (issue 8): 12-14.

Wu, Han, Fei Xiaotong, et al. *Imperial Power and Authority of the Gentry*. Shanghai: Observation Agency, 1937.

Zhang, Yangxi. 'Practical Modern Epidemiology Methods'. *Chinese Journal of Epidemiology* 1994 (issue 1): 315-317.

Zhou, Yongyuan. 'Modern Textile Warp Sizing'. China Textile Warp Sizing Website, http://www.sizing.com.cn/Articles.asp?id=53&lb=th01. Accessed 27 July 2009.

6 Coexistence of Poverty and Cancer

Cheng Pengli and Li Caihong

Abstract

In some relatively undeveloped rural areas of China, cancer is deeply entangled with poverty and other social problems. This study discusses a state-owned farm in Jiangxi, where residents believed that cancer was caused by pollution from a local glass factory and phosphorus fertilizer plant. Our analysis of what the villagers considered 'iron proof' could not confirm an association between cancer and pollution from existing industries, but despite this, residents took their evidence and petitioned various levels of government. Their persistence was fuelled by their resentment of their relative disadvantage in relation to surrounding villages and their desire to get the government to improve their water supply. The case illustrates how different social problems become 'bundled,' and how emotions shape perceptions of risk.

Keywords: rural pollution, water monitoring, risk perception, expert knowledge, social structure, interest groups

Environmental pollution, cancer, and the relationship between them usually fall within the purview of natural science research; but if we look at them within the context of the affected communities, they often become complex social issues. Trends in current research therefore emphasize collaboration between the natural and social sciences in the hope of finding a way to analyse and solve these problems.

Through fieldwork in multiple 'cancer villages', we found that when people feel the incidence of cancer and mortality in their community is high compared to the past or to other communities, and they are unable to find an exact cause, they tend to associate the problem with environmental pollution, and engage in environmental resistance to seek a solution. Because cancer has a high mortality rate, people are very afraid of it, and this,

Chen Ajiang, Cheng Pengli, and Luo Yajuan, *Chinese "Cancer Villages": Rural Development, Environmental Change and Public Health.* Taylor & Francis Group 2020
DOI: 10.5117/9789089647221_CH06

combined with weak regulation of pollution by local government, means they feel environmental protest is justified. At the same time, the relationship between pollution and cancer is extremely complex, and determining a causal relationship between the two is often very difficult. In this situation, the demands residents make of local government and polluting enterprises to solve pollution and reduce the incidence of cancer are often mixed up with their other interests, such as improving living standards or securing other economic benefits. In this situation it is very important for researchers to clarify the real interests that lie behind residents' environmental resistance, in order to avoid being swayed by the emotions of those involved and losing objectivity.

Compared to the prosperous 'cancer village' of Xiqiao discussed in Chapter 5, Jiannan Village is situated in the relatively economically backward province of Jiangxi in central China. Even in this local context, Jiannan has some special characteristics. It is situated on a state-owned farm that was set up on wasteland that was reclaimed for cultivation in 1956. State-owned farms have a special history, but they are mostly now in poor operating condition, and employee incomes are typically very low. Their system of management severely limits their internal economic and social development. Overall, Jiannan's economic situation lags behind that of nearby villages in the region, and compared to Xiqiao, it is a poor 'cancer village'. High rates of cancer not only make matters worse for people who are already impoverished; the disease is intertwined with the problem of poverty. This makes Jiannan residents' environmental resistance even more persistent and intense and affects the community's stability and sustainable development.

Case selection

Two News Reports

Although news reports are less in-depth than academic studies, they provide researchers with quick and effective sources of information. Before starting fieldwork, we found two news articles about Jiannan through internet searches. 'Xin River Pollution, An Unbearable Burden' was published in 2008 in the *China Economic Times*. It described one of the five main rivers that flows into Lake Poyang – the Xin. This river had a history of industrial pollution and was at the time suffering a very heavy toxic load from chemical enterprises established more recently under policies to attract investment. The article described the cancer situation as follows:

> Among the 300 households and nearly a thousand residents living around the former MS factory, there have been continuous cases of cancer in recent years. In total, at least sixty or seventy people have died of cancer, many of whom are of working age, around fifty years old.

The other article, '60 People from 300 Households Die of Cancer' was published in the *Jiannan Metropolis Daily* on 12 December 2005. It described the cancer situation as follows:

> with 300 households and nearly a thousand residents, [...] in recent years there have been continuous cases of cancer, already taking the lives of 60 residents. Since then, the residents' mood changes whenever the subject of cancer is brought up, and they live under the shadow of cancer.

The two articles attributed the high incidence of cancer in Jiannan directly to past and recent industrial pollution. They reported that the cancer mortality rate in the village had reached 600/100,000, which is very high. There were therefore two fundamental questions for us to clarify in our research. Was the cancer mortality rate reported in these articles real? And, if so, was the high incidence of cancer in the village strongly associated with industrial pollution?

Background

Jiannan Village falls under the jurisdiction of S City, in Y County of Jiangxi Province. S City is situated in the south east of Jiangxi, and Jiannan is approximately five kilometres from the county seat. It is located on a state-owned farm called 'Y County Huating General Farm', or 'Huating Farm' for short. In 1984, institutional reform established a township government, but after a few changes, in March 2006, its township status was revoked, and Huating Farm was re-instated. As a result, it is both an agricultural enterprise and a primary level government entity of equal status to a township. The land owned by Huating covers 37 square kilometres, and there are 11,000 permanent inhabitants. Huating administers five branch farms, one village committee, and one community. Jiannan Village is the name we use in this study to refer to the community of approximately 300 households and 1,000 residents who live in the farm employee bungalows around the Huating management office.

State-owned farms are a rather special phenomenon in China. As agricultural economic institutions set up with state investment, they function in a similar way to state-owned industries. According to the available data, in 2007 there were 1,885 state-owned farms in China, with a total workforce of 3.3 million people.[1] The main purpose of establishing state-owned farms in the early years of the PRC (People's Republic of China) was to expand the area under agricultural cultivation and increase the supply of food and other agricultural commodities. Over time, the management system of these farms underwent many reforms, but their fundamental nature did not change. Prior the 1990s, state-owned farms experienced a golden age, and the income of their employees was typically higher than that of other rural residents in the same area. However, as the market economy developed, state-owned farms experienced various management problems, the income of their employees steadily decreased, and the contradictions of their institutional limitations became more pronounced.

Huating Farm, where Jiannan Village is located, was built in 1956. Initially, it was a livestock farm, but in 1959 it switched to agricultural crops. In the 1990s, as state-farms gradually went into decline, some of Huating's workers took the initiative to leave and establish their own businesses. The total number of employees decreased from over 10,000 to 7,000 in 2010. The workers left in the community were mostly old or people with less earning capacity; young, capable people had all left the farm to work or do business outside. Because the farm was not doing well economically, the pension for retired employees was only around 500 yuan per month. Additionally, because the farm's land and buildings were all state-owned, employees did not have land to build their own houses, and they all lived in bungalows assigned by the farm. These residences were mostly built by the collective in the 1960s or 1970s, and they were already very old; in stark contrast to the grander houses built by local farmers on the other side of the road. Overall, compared to the surrounding countryside, Jiannan's development seems to have stalled in the 1970s or 1980s.

Preliminary Findings

The Pollution Situation Reported by Residents

Jiannan is not far from the county town and can be reached in twenty minutes by bus. On arriving in the village, our research team walked into an alley, sat down in front of a house, and started chatting with the woman

1 National Bureau of Statistics of the People's Republic of China, *2008 China Statistical Yearbook*.

who lived there. Because all the people in Jiannan are employees of the farm, and their houses are attached, in no time quite a number of people gathered around us. Everyone was fighting to tell us about the situation in the village, and on hearing that we were there to investigate the issue of pollution and disease, they all had a lot to say.

Even though many people were speaking at the same time, we basically grasped what they were trying to tell us. The enterprise they talked most about, and which was the focus of their concern, was the nearby glass factory. It sat on a small hill near the village. less than a hundred metres from the nearest house in Jiannan, and its tall chimney was visible from the village (see Photograph 17). Residents said that the smoke and dust pollution from the glass factory was severe; clothes they hung out to dry turned black, and tables in their homes were covered daily by a layer of black soot. Some residents said that the well water was so salty there was no need to add salt when cooking. Some of them believed this taste was caused by waste water from the glass factory leaking underground, but a small number of residents also thought it might be due to an explosion at the caustic soda plant in the 1980s, which caused soda water to leach into the ground. When discussing the question of pollution and disease, one woman said:

> There are a lot of people with cancer here. It feels like the number of people dying of cancer started to increase in the 1990s, and this has continued in recent years. This year there were also some that passed away from cancer. The youngest was twenty-nine, and there are many in their thirties and forties.

Some villagers gave us the names of several cancer victims. Two of the women said their husbands had died of cancer, and another old couple said their nephew had contracted cancer the previous year. Some other residents were eager to take us to see the phosphate fertilizer plant, which they said had caused them serious harm. The plant was at the end of the alley, and their houses were right outside the factory walls. As before, a lot of people quickly gathered around us and told us about the problem with the plant. They said the production process was terribly noisy and disturbed them, and the factory chimneys often discharged pungent smoke. Some people said that they had to cover their heads with blankets when they slept. When discussing disease, it seemed that residents were trying to connect illness with the fertilizer plant. One man whose house was right outside the factory said that his wife's 'big stomach' (liver disease) was caused by it and that even the grapefruit tree in their yard was almost dead. They also thought that the Monosodium Glutamate (MSG) factory, caustic soda plant, and soy sauce factory that used to be in the village had polluted the ground water.

The List of Cancer Deaths

In the course of our investigations, we learnt that in recent years Jiannan residents had made numerous petitions over pollution and disease. There were five petition 'activists'. One was Mr. Xu, who used to be an official at Huating farm and had the highest level of education. He had saved materials from previous petitions, including a copy of the cancer death list compiled by the villagers. We went through this list in order, carefully checking each cancer victim's name, age, time of death, type of cancer, and home address. When we mentioned each of the victims' names, Mr. Xu and Mr. Zhou, who were among the five petition leaders, were able to provide the relevant information quickly, which showed that they were very familiar with the people on the list.

Of the 300 households and approximately 1,000 residents of Jiannan, 32 had people who had died of various kinds of cancer over the previous ten years. Mr. Zhou, Mr. Xu and the other three petition leaders had been reporting the pollution problem in the village to higher levels of the government since the beginning of the 1990s and had requested a solution for the residents' drinking water problem. The petition team had counted cancer morbidity and mortality among Jiannan residents in recent years and retained a written copy of the results. Looking at this 'List of Cancer Deaths in Jiannan Village in Recent Years', we can see that as of October 2009, a total of 33 people had died of cancer, of whom 32 died after 2000, and one died before that date (see Table 6.1).

To ensure that we gave an objective and accurate account of the cancer situation in Jiannan, our research team carefully double-checked each cancer death with Mr. Zhou and Mr. Xu, and inquired about the location of the victims' housing, confirming that all 33 people were Jiannan residents. On the basis of Jiannan's population of 1,000 people, we calculated that the annual average cancer mortality rate between 2000 and 2009 was 320/100,000. This was 2.2 times the national rural cancer mortality rate for 2007, demonstrating that Jiannan's situation was abnormal.

Table 6.1 shows that 11 people died from stomach cancer; five from liver cancer; five from lung cancer; four from leukaemia; two from breast cancer; two from colorectal cancer; one from oesophageal cancer; one from brain cancer; one from skin cancer; and one from bladder cancer. The youngest cancer victim was 18, six were between 30 to 40 years old, and 28 were

Table 6.1 Cancer deaths in Jiannan

Serial Number	*Victim Family Name*	*Gender*	*Year of Death*	*Age at Death*	*Type of Cancer*
1	Ling	Male	—	70	Colorectal cancer
2	Zhou	Female	Before 2000	32	Leukaemia
3	Chen	Female	2001	54	Breast cancer
4	Yao	Male	2001	60	Lung cancer
5	Yao	Male	2002	Over 50	Brain cancer
6	Yao's daughter	Female	2001	18	Skin cancer
7	Jiang	Male	2002	32	Liver Cancer
8	Zhu	Male	2002	32	Liver cancer
9	Shen	Male	2002	64	Stomach cancer
10	Luo	Male	2002	62	Colorectal cancer
11	Luo's wife	Female	2004	61	Oesophageal cancer
12	Wu	Male	2002	62	Stomach cancer
13	Zhu	Male	2003	73	Liver cancer
14	Liu	Male	2003	32	Lung cancer
15	Liu	Male	2004	65	Stomach cancer
16	Su	Male	2004	57	Stomach cancer
17	Su's wife	Female	2006	49	Breast cancer
18	Fang	Male	2004	50	Stomach cancer
19	Zheng	Male	2004	54	Stomach cancer
20	Ye	Male	2005	58	Stomach cancer
21	— (Chen's wife)	Female	2005	51	Leukaemia
22	Fang	Male	2005	60	Stomach cancer
23	Fang's son	Male	2009	39	Stomach cancer
24	Huang	Male	2006	50	Stomach cancer
25	Su	Male	2007	62	Lung cancer
26	Zhou	Male	2007	58	Liver cancer
27	Gao	Male	2009	64	Leukaemia
28	Gao's wife	Female	2008	61	Leukaemia
29	Shi	Male	2009	70	Stomach cancer
30	Wu	Male	2009	59	Lung cancer
31	Xu	Male	2009	71	Lung cancer
32	Xu	Male	2009	80	Bladder cancer
33	— (Zhou's wife)	Female	2009	72	Liver cancer

under 70 years old. Six households had two people who died of cancer, including three pairs of husbands and wives, one father and son, one father and daughter, and one mother and son. The types of cancer the villagers contracted were relatively diverse, but liver cancer, stomach cancer, and leukaemia made up the largest proportion. When we were conducting interviews, residents said that it was also common for people to suffer strokes and liver disease. Afterwards, we also conducted interviews in homes and shops on both sides of the road. The respondents were convinced that there had been more than 30 cancer victims over the past ten years. They also said that due to taboos and other moral concerns, locals would definitely not fabricate the existence of cancer victims.

From the above analysis we can see that, at its peak, the incidence of cancer in Jiannan was approximately 2.2 times the national average. However, to determine a scientific link between pollution and cancer, there should usually a relatively strong association between specific pollutants and specific types of cancer, and so analysing rates of mortality from specific cancers in Jiannan is more meaningful than looking at the overall cancer mortality rate. Stomach, liver and lung cancer, and leukaemia were the most common cancers in Jiannan. In 2003, the China Centre for Disease Control (CDC), the National Cancer Research Office, the Cancer Hospital of the Cancer Institute of the Chinese Academy of Medical Sciences and other organizations published the 'Chinese Cancer Control Strategy Research Report', which gave the mortality rate for major types of cancer in certain regions of China between 1991 and 2000. Table 6.2 shows that mortality from stomach cancer in Jiannan was approximately 4.88 times the average national rural level, and liver cancer and lung cancer were 2.22 and 3.5 times as high. It is especially noteworthy that in the space of ten years, three people in Jiannan died of leukaemia, which is nearly ten times the national rural average.

Table 6.2 Mortality rates for major cancers nationwide and in Jiannan (1/100,000)

Type of Cancer	*Urban*		*Rural*		*Jiannan*	
	1991	*2000*	*1991*	*2000*	*Number of deaths (people)*	*Mortality rate*
Total	123.92	146.61	101.39	112.57	32	320
Nasopharyngeal	2.17	2.14	1.70	1.78		
Oesophageal cancer	8.94	9.06	16.32	15.24	1	10
Stomach cancer	19.69	19.94	22.55	20.94	11	110
Colorectal cancer	8.53	11.19	5.14	6.04	1	10
Liver cancer	19.63	22.19	22.25	26.06	5	50
Lung cancer	32.53	42.08	14.29	21.11	5	50
Breast cancer	3.11	4.32	1.51	2.11	2	20
Leukaemia	3.60	3.82	3.31	3.18	3	30
Cervical cancer	1.49	1.08	2.31	1.91		

In-depth Investigation

Emotional Residents

The selection of a research site is crucial in the social sciences. When we first arrived in Jiannan, we heard from residents that pollution in the village was very severe, and that in addition to the current problem of emissions from local industries there had also been a history of pollution. Looking at the list compiled by the petition leaders, it seemed that the incidence of cancer was relatively high, and there were some cancer patients who contracted cancer when they were very young, as well as one household in which multiple people had died of cancer. All these factors suggested that this was the perfect place to conduct our investigation into the relationship between pollution and cancer.

Social science requires objectivity. Researchers can make their choice of topic, but in the process of research they must do their best to remain unbiased and neutral. This is not easy. In the initial stage of research in Jiannan, we were influenced to some extent by the emotions of the villagers. As strangers, we were not familiar with the situation of the community, and their strong feelings affected us. We felt sympathy for them. However, we later learnt that Jiannan residents had started petitioning in the early 1990s, and so relating their grievances to strangers was already effortless for them. Although from the start we made it clear that we were only there to do research, and that we were not reporters, the residents still hoped that we would help them relay their story to the world outside.

Mr. Zhou was the most active of the petition leaders. He was over 60 years old and a retired employee of the farm. He had never been to school and was basically illiterate. Mr. Zhou was the most enthusiastic and most dedicated activist. Starting in about 1992, he led residents to petition at government departments at the county, city, and provincial level. He himself and other people in Jiannan told us that his son and various relatives worked for the government, and some of them held quite high positions. Villagers believed that it was because Zhou had these connections that the public security police did not dare to bother him, and that is why he was not afraid of anything.

During our fieldwork in Jiannan, Mr. Zhou always accompanied us to different houses to conduct interviews and collect materials. As we got to know him, we became aware, on the one hand, of his perseverance in petitioning and, on the other, of how stubborn he was in his perception of the relationship between pollution and cancer. Zhou insisted that so many residents were dying of cancer because of current pollution from the glass factory. He considered the findings of tests of water samples that had been conducted by the authorities some years earlier were effective evidence of pollution, and when we attempted to explain the meaning of the water quality indicators used in the tests, and the fact that they did not confirm a relationship to pollution from the glass factory, he expressed strong displeasure, asking:

> Were you sent by the environmental protection bureau? If you were sent by the bureau, I will not speak with you any longer.

In fact, Zhou's persistent petitioning and stubborn belief in the relationship between pollution and cancer may have been related to his personal life experience. Zhou's (second) wife had passed away from liver cancer a year before at the age of 71, and his stepson had passed away seven or eight years before, also from liver cancer. Two family members dying in succession must have been very painful for him and in his sub-consciousness, petitioning was perhaps also a way to seek justice for his late wife and son. Furthermore, Zhou's home was right outside the walls of the glass factory, and the ash from its chimneys directly affected his home. But although we sympathized with him, our research showed that his conclusion about the cause of cancer was by no means certain.

Pollution from Two Eras

Industrial pollution in Jiannan was in fact from two periods. There was legacy pollution from past emissions and current pollution from relatively new enterprises. From the 1970s to the beginning of the 1990s, Huating Farm established a group of collective enterprises, including a soy sauce factory, an MSG factory, a brewery, a caustic soda plant, and a glass factory. The food processing enterprises were large scale and brought in a good revenue at the time. Because Jiannan is situated at the centre of Huating's operations, these factories were all very close to the village (see Table 6.3).

The soy sauce factory was the first to be established and taking advantage of being able to produce its own raw materials, Huating went on to build the Huating MSG Factory in about 1970. The enterprise grew steadily and Huating MSG soon became a well-known brand. For a long time, both the soy sauce and the MSG factories made good profits. However, they also became quite major sources of pollution, discharging a huge amount of waste water that not only affected the local community, but also nearby rivers. The caustic soda plant was built in 1980, producing for industrial use. Not long after the factory was built there was a boiler explosion that resulted in a several tons of caustic soda seeping into the ground and heavily polluting the river and ground water. The first glass factory was built in 1987 and emitted waste gas and water. It shut down in 1995. In 1997 that factory and its equipment were rented to a private entrepreneur from Fujian (see below).

Table 6.3 Huating industrial enterprises from the 1970s to the 1990s

Factory Name	*Date of Establishment*	*Date of Closure*	*Pollution Situation*
Soy sauce factory	1970s	Beginning of 1990s	Waste water
MSG factory	1970s	Beginning of 1990s	Waste water
Caustic soda plant	1980s	Two years after establishment	Several tons of caustic soda entered the ground
Old glass factory	1987	1995	Waste gas (black smoke), waste water

As in the nation as a whole, after the glory days of the 1980s, collectively-owned enterprises in Jiannan began to decline in the early 1990s. This was due partly to their own limitations and partly to the broader impact of the market economy. Many enterprises underwent reform or were closed. In the middle of the 1990s, almost all of Huating's collectively owned enterprises sank into difficulties, and one by one they were reformed or shut down. By around 1995, pollution emissions from the enterprises listed in Table 6.3 had basically ceased.

However, this does not mean that pollution from these industries ceased to present a health risk. When we look at the location of Jiannan village in relation to the factories, which is shown in Table 6.3, Jiannan is low lying, while the surrounding terrain on which the factories were located is relatively high. Effectively, Jiannan is positioned at the bottom of a funnel. We can therefore speculate that the pollution discharged by those factories may have affected the ground water which Jiannan residents had been drinking for a long time, even after the factories were closed.

Huating never resumed industrial activity after the decline of its collective enterprises in the mid-1990s. At the time of our research, there were two main industrial enterprises near Jiannan: the glass factory, and the phosphate fertilizer plant. The old glass factory and its equipment had been rented to a private entrepreneur from Fujian Province in 1997. Originally, the factory used metal ores to produce glass products. After the new owner took over, he continued to use the original equipment, but he bought old glass bottles and melted them down and processed them into new glass bottles for safflower oil. The new glass factory produced some waste gas, mostly in the form of smoke dust from burning coal as the main fuel. A small amount of waste water was also produced and stored in a large pool in a corner of the factory compound. The phosphate fertilizer plant also used to be one of Huating's collective enterprises and was later contracted out to a private owner. The new operation mixed phosphate rocks bought on the market with other fertilizers. The production process did not involve any complex chemical reactions or discharge any new pollutants. However, residents near the plant reported that when it was in production, it was quite noisy.

Residents' Views on the Cause of Disease

Not only the five petition leaders but also ordinary residents of Jiannan believed that the high incidence of cancer in the village was caused by pollution, and that the existing glass factory was the main culprit. There are several rationales for their thinking this despite the fact that we found it difficult to find evidence of a relationship between the cancer and pollution from the glass factory or the phosphate fertilizer plant. First, residents pointed to what they referred to as 'iron proof' – an official report stating that Jiannan's ground water did not meet quality standards. Second, although residents were aware that historical pollution from the MSG factory and the caustic soda plant had polluted the ground water, these factories no longer existed and so to attribute cancer to this historical pollution would be futile from the point of view of seeking redress. The two factories that were still in operation therefore naturally became the prime suspects. Lastly, water is an important medium through which pollution causes harm to human health. Since the 1990s, Jiannan villagers had noticed that their well water had a peculiar taste and demanding a change in their water source was therefore always a major component in their repeated petitions.

From the end of 1999 to the beginning of 2000, residents' representatives, the local government and the Health and Epidemic Prevention Station took water samples from three residential households in Jiannan and sent them to the county and provincial health and epidemic prevention stations for laboratory testing. Table 6.4 shows the results.

The test results show that the water quality did not meet the standard for some indicators. However, this only shows that the well water in Jiannan was polluted to some extent; no specific carcinogens were found. It is therefore very difficult to prove on the basis of this test report that the high incidence of cancer among residents had a definite relationship with well

Table 6.4 Test results from the three water samples from Jiannan

Index	*Water sample 1 (Liu household)*	*Water sample 2 (Liu household)*	*Water sample 3 (Li household)*	*National standard limit (2006)*
Chroma	*20*	<5	10	15
Turbidity	*5*	*6.5*	*27*	1
Smell and taste		*Astringent*	*Astringent*	No odd odours or taste
PH value	*9*	*13*	6.77	No less than 6.77 and no more than 8.5
Total hardness	150 mg/l	*620.6mg/l*	*560.5 mg/l*	450mg/l
Ammonia Nitrogen	–	0.8mg/l	*1.6mg/l*	0.5mg/l
Iron	<0.3 mg/l	0.03 mg/l	*0.31 mg/l*	0.3mg/l
Manganese	*0.105mg/l*	0.06mg/l	*1.69mg/l*	0.1mg/l
Copper	<1.0mg/l	0.01mg/l	0.005mg/l	1.0mg/l
Zinc	<1.0mg/l	0.005mg/l	*1.08mg/l*	1.0mg/l
Lead	<0.05mg/l	*0.035mg/l*	*0.02mg/l*	0.01mg/l
Chromium	<0.05mg/l	<0.04mg/l	<0.004mg/l	0.05mg/l
Cadmium	–	*0.05mg/l*	*0.006mg/l*	0.005mg/l
Arsenic	<0.05mg/l	<0.05mg/l	<0.05mg/l	0.01mg/l
Cyanide	<0.05mg/l	<0.002mg/l	0.002mg/l	0.05mg/l
Chloride	–	*290mg/l*	*460mg/l*	250mg/l

Notes:
1. Between 1999 and 2000 Jiannan residents took three different water samples from three different wells in homes in the community and sent them for tests. The testing institute for water sample 1 was the County Health and Epidemic Prevention Station. Residents thought that these test results were unreliable, and they then took water samples 2 and 3 to the Provincial Health and Epidemic Prevention Station. The national standard was added by the author for comparison.
2. Italics indicate values exceeding the national standard.

water pollution. Nonetheless, some residents believed that since the official water tests showed problems with the water quality, the high incidence of cancer must have been caused by their drinking this water over a long period of time. Interpreting the test results required a level of specialized knowledge that the villagers did not have, and so in their hands, results that showed 'problems' with the well water quality became 'iron proof' that the water was the cause of cancer.

Statements from Other Institutions

After gaining an understanding of residents' attitudes towards 'pollution and cancer', we visited the Deputy Director of Huating Farm in charge of industry. He explained the situation as follows. First, the existing glass factory produced new glass bottles by melting down old ones at a high temperature. Emissions from this process were mainly conventional pollutants, namely sulphide and ash produced in the process of burning coal as fuel. Just to deal with the problem of ash, the glass factory had already invested between one and two million yuan in technological upgrades, and the situation had improved compared to the past, although sometimes stokers adding coal unevenly caused black smoke to rise suddenly. As for the arsenic (arsenic trioxide) which villagers believed the glass factory used in the production process, the Deputy Director said that the old factory did use this substance, but that the current operation no longer did because there was no need for it. As for the phosphate fertilizer plant, it mixed together ready-made phosphates bought in from outside and did not produce any pollution. It only emitted a small amount of dust, and there was very little waste water. The Deputy Director believed that it was fundamentally true that there was a high incidence of cancer in Jiannan, but he could not clearly determine the cause. As for the residents' petitions, he considered that they were 'making a fuss', about pollution mostly because they thought it would bring them other benefits.

To understand the pollution situation further, we visited the County Environmental Protection Bureau (EPB). Employees from the Pollution Control Division admitted that the Bureau had received reports from Jiannan residents about pollution in the area, and they had conducted investigations to verify the situation. But they indicated that the pollution in Jiannan was not serious. The existing glass factory melted old glass, and it was mainly an issue of gas emissions. The local government had already planned the relocation of the glass factory, and it would soon be moved to a county industrial park. The EPB officials confirmed that the fertilizer plant mainly processed phosphate powder, and that it discharged a certain amount of dust, but there were no chemical reactions involved in the production process, and waste water was almost non-existent. We also discussed the test results of water samples from Jiannan with an EPB staff person, who told us that it was common in the area for manganese levels in ground water to exceed standards. In his opinion, there was no scientific proof to link the high incidence of cancer among residents with pollution.

After the EPB, we continued our investigations by visiting the county CDC. Staff of the Division of Local Chronic Diseases expressed doubts about the list of cancer victims provided by the residents. They did not think that cancer in Jiannan displayed a pattern and considered that the incidence was not much different from other places. In response to complaints from Jiannan villagers, the county CDC had also arranged a professional epidemiological investigation in the area. From 2007 to 2008, they tested ground water drinking sources in Jiannan, taking samples at multiple points, including downstream, and from both deep and shallow water. Aside from the excess levels of cadmium and manganese that were present for geological reasons, the water quality did not have any other problems. As for the thirteen or fourteen cancer cases reported by residents, CDC staff said that they once interviewed more than 100 households accompanied by township officials, and confirmed only half of these, or about six or seven cases. They said the main reason for the discrepancy was that cancer diagnoses given in small clinics could not be counted as formal statistical data, and that some residents might also have attributed deaths caused by other diseases to cancer.

As for the reasons behind the incidence of cancer, the CDC staff thought that it would be very difficult to determine the cause. They explained that the county CDC had limited testing ability, and there were many situations that they could not explain. For example, determining the cause of stomach cancer would require monitoring all kinds of variables that might affect the digestive system. In terms of diet, Huating Farm was already similar to the wider society, and there were no particularities in terms of products residents bought in from outside, so there was no way to make a meaningful comparison to other communities. Furthermore, because residents ate a large variety of foods on a daily basis, it would be difficult to decide which to test. There was also at that time no way to test for aflatoxin [a known carcinogen] in food. The residents of Huating originated from many different places and there was nothing special about their lifestyle habits. After eliminating water and food-related causes, they mentioned that Hepatitis B is one of the factors that should be considered as triggering the onset of cancer. But Hepatitis B is prevalent all over China, and Jiannan is not unusual in this respect either. Overall, the history of the farm was similar to that of other such farms in Jiangxi Province. They were all cultivated on wasteland, and did not have any history of mining, thus eliminating the possibility of heavy metal pollution.

Even though the results of the investigation by the CDC could confirm neither the high incidence of cancer in Jiannan nor the cause of cancer, it should be noted that the scope of the CDC investigation was the entire town, which had a relatively large population, while we had learned that the high incidence of cancer was mainly concentrated in the individual village of Jiannan. Furthermore, we do not know the details of the methods used in the CDC's investigation. However, CDC staff reminded us that the information provided by Jiannan residents could be suspected of exaggeration, because there were emotional factors involved. They said that because most of the villagers were laid off or retired employees of the farm, there were difficulties with pension payments, and their living conditions were relatively poor. They might therefore be exaggerating the facts in order to support other demands.

Overall Analysis

At the start of our research we were influenced by the villagers' emotions and generally believed what they said about the connection between pollution and cancer. But after examining the water test reports and discussing with staff of the government agencies concerned, our thinking gradually became clearer. Even though vulnerable people deserve sympathy, as researchers, we should always make an objective evaluation of the facts. After an extensive investigation, an integrated analysis of the full range of information we had obtained, and extensive discussion within our research team, we came to the following conclusions.

First, there was a certain level of pollution in Jiannan. Waste water discharged by the MSG factory over a long period, and caustic soda from the boiler explosion in the collective era may have entered the ground water and polluted well water in residents' homes. The high pH and turbidity values in the water test results shown in Table 6.4 may have been related to this. However, pollution from the new glass and fertilizer industries was minimal. Residents believed that the glass factory used arsenic, but the water test results showed no irregularities in the arsenic content.

Second, the villagers' claims regarding the high incidence of cancer were basically correct. After verifying the information they provided about cancer deaths in Jiannan, we came to believe that there was indeed a high incidence of cancer in the village. Because the disease has a high mortality rate, rural people consider contracting cancer to be a disastrous and disgraceful thing, and they would not usually make it up. Meanwhile, the CDC's research was influenced by all kinds of factors and may not accurately reflect the situation of cancer deaths in that particular community.

Third, the residents' view of the relationship between pollution and cancer was difficult to confirm, and the cause of the high incidence of cancer was hard to determine. The water tests that residents considered to be 'iron proof' could not support their position that there was a clear relationship between pollution and cancer. The tests showed excessive levels of manganese, but existing scientific research does not support a connection between manganese and cancer. Cadmium does have a relatively strong association with cancer, but the enterprises operating in Jiannan at that time did not involve substances containing cadmium, and one of the two water samples was close to normal on this indicator. As for where the manganese in the water samples came from, County EPB staff believed that the red soil in the area contains relatively high levels of iron, and manganese and iron generally coexist. Manganese often exceeds the standard in the area and so these high levels were very likely the result of the region's underlying geology.

Residents' Resistance and Motivations

From the late 1980s, Jiannan residents started to notice that their well water did not taste right. They thought it might have been polluted by waste water discharged by collective enterprises and be harmful to their health. Starting in 1992, they began to petition, and they continued to do so for more than ten years. The main purpose of the petitions was to demand that tap water be installed for the community to provide safe drinking water. Later, in 1995, 1999 and 2000, Jiannan residents sent representatives to all levels of government to petition, but with minimal effect. After 2000, as more community residents died of cancer, the frequency of petitions also increased, and they appealed to higher and higher levels of government, including the People's Congress and the EPB, at the county, city and provincial levels. As the result of these efforts, in 2009, the local government took advantage of nearby road repairs and connected tap water pipes to Jiannan residents' homes. They were finally able to drink running water.

Over the course of more than ten years of petitioning, Jiannan residents gradually formed a petition 'leading group' with five core members. Of these, Mr. Zhou was the most enthusiastic, even though he had never attended school. Zhou believed that the installation of tap water was due to the group's active petitioning, that because of their persistent efforts some high-level leader had finally paid attention to the problem, and they were able to get tap water. Although it is hard to verify the details, Mr. Zhou was convinced that petitioning had been effective. During the many years that the petitioning went on, Mr. Zhou had also come under heavy pressure. For example, officials leaned on his family members to persuade him to give up petitioning. He felt that their success had been due to the scale of the protests.

> You have to have a lot of people, a few hundred people causing a ruckus in the county, see if they solve the issue or not. If you just go with a few people, they won't bother with you.

Residents solved the drinking water problem through petitioning and fund-raising, but recently there had been more news that made them feel aggrieved. The county government had built a waste water treatment plant south of the town and demanded that Jiannan residents pay a water treatment fee of 0.8yuan/ton. The villagers believed they were the victims of pollution, and that waste water from the factories had polluted the underground water, making it impossible for them to drink their own well water. Given this, they felt it was unfair that they should have to pay first for tap water and now for waste water treatment. The county government's rationale was that Jiannan now belongs to the urban area and so they must pay for water treatment.[2] Mr. Zhou told us that the residents thought this fee was unreasonable, and that they would petition about it in future. This episode again reveals how people's different positions and considerations shape their views about what is reasonable and what justifies 'making a fuss.'

Demands for Remediation of Pollution

Aside from demanding that the tap water issue be solved, Jiannan residents also actively sought to resolve the problem of existing pollution. Because they believed that the high incidence of cancer in the village was directly related to the glass factory, they tried to negotiate with the enterprise and caused multiple conflicts. These were particularly severe in 2010. In September of that year, more than 50 residents of Jiannan went to the glass factory with Mr. Zhou as their representative. They locked the factory's front gate, preventing cars from entering or leaving. Afterwards, the farm director came out and promised to resolve the problem after National Day, and only then did the residents unlock the gate. But after National Day, nothing happened. Mr. Zhou went to find the factory director, but both he and the Party Secretary hid and would not see him, so the villagers went back to the factory. After stalling for a very long time, the factory directors agreed to offer only 50 yuan to each person who had come to protest. The residents did not accept this offer. They demanded that the County EPB order the glass factory to change its production equipment and add pollution mitigation equipment, and even that the EPB should close the factory down.

In November 2009, according to 'Y County 2009 Special Environmental Protection Action Plan for the Remediation of Enterprises Illegal Waste

2 Water is free in rural areas, while urbanites must pay a fee based on the amount of water used.

Water Discharge to Protect People's Health', the county government issued a notice to the glass factory notifying it of the deadline for remediation. Three enterprises, including the glass factory, were put under deadlines for pollution control. The pollutants targeted for control in the case of the glass factory were dust and sulphur dioxide. (From this we can determine that these were the emissions that exceeded the standard). As for the waste water problem reported by residents, we do not know whether or not these emissions were over the legal limit). The government also requested the EPB, and the electricity, business, trade, and other sectors to ensure proper supervision of the factory according to regulations. The deadline was set for 31 May 2010. According to the 'Environmental Protection Law of the People's Republic of China', if the glass factory did not complete the tasks by the deadline, in addition to collecting an additional fee for excess emissions, it would be possible to impose a fine that would be set in relation to the harmful consequences, or to order the factory to suspend production or be shut down. However, the order to suspend production or shut down a factory is made by the county government which issued the deadline, whereas fines are decided by the county EPB. The EPB does not have the power to decide whether to shut down an enterprise or suspend production.

The Motivation behind Residents' Continuous Petitioning

In our interviews with Jiannan residents we found that the five-person petition group and a number of other residents were very active about petitioning. We were curious about how Jiannan villagers had been able to maintain this continuous level of petitioning from 1992 until 2009, and what drove them to do so. We came to the following conclusions.

First, there were the problems of pollution and of disease. There was a certain level of pollution in Jiannan. Waste water discharged by the Huating General Farm during the collective enterprise era had entered the ground water, and after many years it may have seeped into the water in the villager's wells. The ash currently discharged by the glass factory and the phosphate fertilizer plant also caused a certain level of air pollution. At the same time, the cancer mortality rate in the community was so high that it made residents feel they were facing an enormous health risk. In the space of ten years, 32 people died from various kinds of cancer, and this threatened every aspect of life in the community, which was constantly shrouded by the shadow of death. During fieldwork, we found that Jiannan villagers were very close to each other, but in households where multiple people had died over the course of a few years, the remaining family members were forced to leave their homes and seek refuge in other places with friends or family out of fear that the same fate would befall them. Given that they felt they faced such a huge health risk, villagers' desire to eliminate the pollution sources was a totally normal reaction.

But there were also conflicts of interest involved in this environmental struggle. Before the 1990s, the farm's enterprises had high output, workers' incomes were relatively high, and their living standards were good. Mr. Zhou told us that the 1980s were the best period for the farm workers, especially around 1984 and 1985, when their monthly bonus could be as high as 300 yuan, and the basic wage was only around 40 yuan. Compared to the national average annual income of 1,148 yuan in 1985 (equivalent to 95.6 yuan/month),[3] the income of workers at Jiannan farm was very high indeed. In addition to their salaries, they also received 25 kilograms of food rations. Due to declining economic revenues and institutional changes, the financial situation of the farm began to deteriorate in the 1990s. Compared both to the past and to the residents of surrounding villages, the workers came to feel poor, left behind, lost, and off-balance. They were employees of a state-owned enterprise, but their income was lower than that of ordinary farmers; and even though they lived in housing allocated by the collective, their 1970s homes were already very out of date compared with new houses being built in the countryside. They did have stable employment, but they were still engaged in agricultural labour, and had limited opportunities for professional development. As all the disadvantages of the state-owned farm became clearer, the dissatisfaction of the workers mounted. There is therefore no way that the residents' various struggles to demand pollution control and reduce health risks were not intertwined with other interest demands.

The Deputy Director of the farm believed that residents' petitioning was not only about environmental pollution and cancer, but that they also had ulterior motives. The residents complained partly because they hoped that the glass factory could be moved away, but they also hoped to receive financial compensation. The glass factory was rented by the farm to a private party, and so although the farm got a certain amount of rent, the residents themselves did not get any obvious economic benefits. In the meantime, the township government had also reported multiple times, both in written form and orally, to the county government, asking that the glass factory be relocated to the county industrial park. When our research concluded, the

3 Liu and Nie, 'China Wage Income Tax Rate Research', pp. 59-67, 207-208.

new site of the glass factory had already been selected and construction was underway in the industrial park. The factory was expected to move out of the residential area in the near future.

Works Cited

Liu, Yi, and Nie Haifeng. 'China Wage Income Tax Rate Research'. *China Social Sciences* 6 (2005): 59-67, 207-208.

National Bureau of Statistics of the People's Republic of China. *2008 China Statistical Yearbook*. Beijing: China Statistics Press, 2008.

7 Problematization and De-stigmatization

Li Caihong and Cheng Pengli

Abstract

Although they live up and downstream on the same polluted river, residents of the two villages of Liangqiao and Shangba have quite different perceptions of the risk of heavy metal pollution from a local mine and different responses to it. At the time of this research, Liangqiao villagers were actively trying to problematize pollution, and attract attention in order to resolve it, while Shangba had already gone through this process of problematization, and realized that although it had brought some relief, being known as a 'cancer village' also had downsides. It was now trying to 'de-stigmatize' the community. The differences between the two communities were largely related to factors such as clan organization, local or migrant origin, population size, economic power and other such factors.

Keywords: heavy metal pollution, media, village elites, village scale, comparative research

Introduction

The villages in this case study are situated in the mountains of northern Guangdong province where some large mining enterprises are located. Mine tailings and waste water flow down the course of the river, and over many years they polluted the villages mid and downstream. The farmland around the villages suffered heavy metal pollution, which damaged villagers' health and also led to secondary social problems.

The Mine

B mine is located in a county in northern Guangdong Province. The county is surrounded by mountains which rise in waves from the northeast to

Chen Ajiang, Cheng Pengli, and Luo Yajuan, *Chinese "Cancer Villages": Rural Development, Environmental Change and Public Health*. Taylor & Francis Group 2020
DOI: 10.5117/9789089647221_CH07

the southwest and cover around 80% of the county. Due to the hilly and uneven terrain, there is not much room for paddy fields. The main food crop is rice and the main cash crop is sugar cane. The topography of the county is diverse, the geological structure is complex, and the area is rich in mineral resources.

B mine exploits the large deposits of multiple metals in the area. The main minerals are iron ores but there is also copper, tungsten, antimony, molybdenum, gold, silver, and other non-ferrous metals. Founded in 1958, B mine is a provincial state-owned mining company. It has always used opencast mining, and over time the tailings and waste water flowed down along the river, polluting the water sources and soil of the villages along its banks with heavy metals. Close to B mine there were also some relatively small-scale municipal and county mining companies. Over a certain period, lack of sound management meant that some small private enterprises also mined secretly in the area. Because these small enterprises were focused on economic efficiency, their negative impact on the environment was even greater.

The River

Near B mine are two gorges that channel the Li River and the La River. The two rivers merge about ten kilometres downstream, forming one larger river, the Heng. The point where the Li and La merge is where Liangqiao, the first village downstream of B mine, is located. A few kilometres further downstream of Liangqiao is a larger village called Shangba (see Figure 7.1).

The Heng River originates in Liangqiao, runs through more than ten administrative villages, merges with larger rivers in the county, and then, after several turns, finally enters the Pearl River. Because the water of the Heng was affected by pollution from the mine tailings, at the time of our research it always looked a brownish yellow, mainly due to the high concentration of ferrous ions. As they flowed along the river, the tailings and silt carried by the water were gradually deposited, causing a thick brown sludge to build up on the upstream riverbed of the Heng. Even though B mine built a tailing dam upstream of the Li and La, because the damn was very old there was a serious build-up of silt in it, and the water often overflowed and ran downstream.

The Villages

Liangqiao and Shangba are both relatively close to B mine, and they were the villages most seriously affected by the pollution. However, the situation in

Figure 7.1 Schematic diagram of the mine, the rivers and the villages

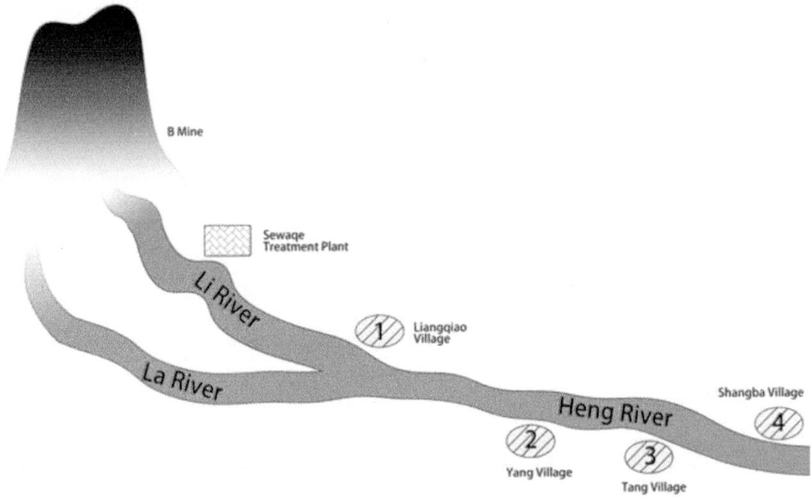

the two villages is quite different. Liangqiao is located at the highest point of the Heng River where the Li and La merge. It is the village closest to the mine and was also the most heavily polluted. Even though Liangqiao is an administrative village, it is very small. There are only four village groups with a total of 90 households and a population of 382 people. This population would usually only make up one or two groups in a normal Chinese village. The Liangqiao Village Committee consisted of four people: the posts of Village Head and Party Secretary were held by the same person, there were two other committee members, and an additional cadre in charge of family planning.

Liangqiao was gradually formed after the founding of the People's Republic of China (PRC). In the 1950s, in response to the national policy, the local government organized some villagers to relocate to places around Liangqiao to cultivate wasteland, and this population gradually grew into three villager groups. In the 1970s, other people were relocated from Henan, Hunan, Jiangxi, and other places outside the province, and gradually formed another villager group. Even at the time of our study, villagers who came from outside the province could not speak the local dialect properly. Liangqiao is situated upstream of the Heng River, and its natural geographical conditions are rather poor, with few paddy fields and not much arable land. The economic situation of Liangqiao is worse than that of Shangba.

Shangba is located downstream of Liangqiao. It has relatively flat terrain and covers a large area. Because of the heavy pollution in the village and the large number of people dying of cancer, a lot of journalists and researchers visited Shangba, and with so many news reports and research results circulating, the village became 'famous'. Compared to tiny Liangqiao, Shangba can be considered a 'giant village'. At the time of our research it had 21 villager groups and a population of more than 3,400 people. Compared to the complex origins of the residents of Liangqiao, Shangba villagers had mostly all lived there for generations, and have the same ancestral name, He. The clan had an ancestral hall. As in Liangqiao, Shangba villagers mainly planted rice and sugar cane but they had more arable land per household, the land was more fertile, and water sources were more abundant and convenient.

Selection of the two villages

Before conducting fieldwork, we had preselected Shangba as our research site. However, during the course of our investigations we discovered that not only was Liangqiao also a victim of mining pollution, but the pollution was actually more severe than in Shangba. Because the population of the village was smaller, Liangqiao never received much attention from outsiders, but it offers an interesting comparison to Shangba, enriching the analysis. We therefore included it in as an additional case study.

According to the media reports we read before starting our fieldwork, Shangba had suffered from pollution caused by B mine for many years, there was a high incidence of cancer among local villagers, and the village had already attracted a lot of attention from experts, the media, NGOs, and other organizations. It was a famous 'cancer village' in China.

Our experience from previous studies of 'cancer villages' was that, as victims, village committee cadres were usually willing to talk about the situation of pollution and disease with researchers. So we were surprised when we arrived at the Shangba Village Committee and the committee staff told us that in order to conduct our study, we must first apply to the town government for approval, and only then could the village committee receive us. We went to the town government, and after an official there called them, the Shangba Village Committee agreed for us to return for research the next day. Surprisingly, the town government official volunteered to arrange for us to visit the waste water treatment plant, and contacted Ms. Guo, a staff person on the Village Committee of Liangqiao, which is near the treatment plant.

We speculated that the main aim of the town government in arranging for us to visit the treatment plant was to make us aware of the efforts the government had made to deal with pollution from the mine. The treatment plant was only a little more than ten minutes away from Liangqiao. It was built with funding from B mine and mainly treated waste water from tailings upstream of the mine. During our visit, the plant was still under construction. (See Photographs 18 and 19). Ms. Guo told us that the plant could not fundamentally solve the pollution problem, because the quantity of water upstream was too great, and the capacity of the plant was limited. Moreover, this plant was built downstream of the Li – one of two rivers upstream of the Heng – and waste water from the other river, the La, was still untreated.

Visiting the waste water treatment plant also gave us a surprising opportunity to learn about Liangqiao. During our visit, Ms. Guo told us that even though Shangba was the media famous 'cancer village', in terms of pollution and cancer, the situation in Liangqiao was even worse. Liangqiao is at the uppermost part of Heng River, where the Li and La Rivers merge. It is therefore the closest village to the mine, and the first village along the Heng to be affected by mine tailings and waste water. Down from where the Heng starts, the riverbed and riverbanks gradually widen, and the terrain becomes flatter. There are two villages between Liangqiao and Shangba that are about three to five kilometres away from each other as the crow flies. Ms. Guo believed that of these villages along the Heng, Liangqiao was the most heavily polluted because it is upstream.

There are two main kinds of damage pollution can cause to agriculture. The first is tailings from upstream that settle directly on paddy fields and dry land, covering the original soil and making arable land impossible to cultivate. The other is the heavy metals contained in tailing water. If this water is used to irrigate agricultural fields, the crops will be contaminated. Pollution had also had a severe influence on the lives of villagers. They were forced to give up drinking water from the village well years ago and used their own funds and labour to get mountain spring water from a high mountain with plastic piping. But the volume was limited and only enough for daily use, not for irrigating farmland.

Why, given that the pollution in Liangqiao was worse than in Shangba, were there so few reports in the media? Ms. Guo believed that the excessive attention of the media to Shangba has benefited the village in solving its pollution problem, and that this was a little unfair to Liangqiao. She felt that Liangqiao had been neglected mainly for the following reasons.

First, Liangqiao's collective power was weaker. Not only did Liangqiao have a small population – it was a classic 'micro village' – but the composition of its population was also very complex because the villagers all had moved there from elsewhere, and some even came from outside the province. In contrast, Shangba was basically comprised entirely of locals, and most of them belonged to the same ancestral clan. This meant that when engaging in environmental resistance, there was a big difference in power between the two villages and this led to very different outcomes. Second, the economic situation in Liangqiao was bad. Due to its geographical position, income from agriculture was limited. A lot of young people migrated for work, but the village had produced very few 'big shots'. In comparison, Shangba always had someone from the He clan in the town or higher levels of government to speak for them when needed. Lastly, because it was not famous, government funding and outside aid usually did not reach Liangqiao. For example, at the time of our research, Shangba had recently managed to get more than 1.4 million yuan in joint funding from B mine and the local government to build a water reservoir and install tap water, whereas Liangqiao only got a pathetically small 'pollution remediation fee' every year.

Pollution and Cancer

Overview of Pollution in the Villages

From the 1970s, mining waste water severely polluted villages downstream of B mine. According to reports, water from the tailings dam had an acidity level more than 1,000 times the national standard.[1] At the time of our research, the water of the Heng was a reddish-brown and very murky. It should no longer have been used to irrigate farmland. But because Liangqiao is on high ground and has no alternative water sources, the villagers were still using water from the Heng to irrigate their fields. Ms. Guo provided a copy of a soil pollution monitoring report on Liangqiao, which showed that the cadmium content in the soil was 2.5mg/kg, more than eight times the national environmental standard for soil (0.3mg/kg). Because water sources and soil were polluted by the mine, the heavy metal content in crops also severely exceeded the standard. Liangqiao's arable land was limited to begin with, and pollution of the paddy fields made life even more difficult for villagers. When the research team interviewed village committee member Ms. Guo in her home, the bleak and desolate state of her house was surprising.

Even though Shangba is in the mid to downstream section of the Heng, it still suffered from severe pollution. By the time the Heng reaches Shangba,

1 Southern Metropolis, 'Guangdong Wengyuan Cancer Village Research: 250 People Died Due to Pollution' .

it has already been diluted by water from other tributaries. The colour was markedly lighter than that in Liangqiao and the water clearer, but the pollution was still serious. Statistics from the Second National Agricultural Soil Survey showed that Shangba's polluted farmlands contained levels of lead 44 times higher than the national standard. Cadmium content exceeded the standard by 12 times, and the pH level was around 4.[2] An expert from a university in Guangdong Province conducted testing of the crops and vegetables in Shangba multiple times and concluded that the acidic mineral water discharged by B mine had already caused severe contamination of crops, vegetables, and fruit, with cadmium pollution the most prominent. The cadmium content in sugar cane, bananas, lettuce, bitter gourds, eggplant, peppers, water spinach, sweet potato leaves, and rice were 149, 187, 7.7-29, 6.6-10.5, 15-24, 7, 15-59, 33, and 2 to 5.7 times the standard, respectively. The lead, manganese, iron, copper, tin, and cadmium content in the Heng were 11, 12, 224, 6.6, 3.7, and 10 times the levels in water not affected by mine waste water.[3]

Long before these tests were conducted by authoritative institutions, ordinary villagers had felt the impact of pollution from the mine through their daily life experiences. Elderly villagers told us that B mine started operations in around the 1970s and they saw the river water starting to turn red around 1973. If their hands touched the water, they would get blisters, and eating fish from the river caused diarrhoea and vomiting. Later, fish and apple snails gradually disappeared. Many villagers believed that through irrigation, the river water also contaminated the soil and crops. Additionally, the wells in village were typically only two to three metres deep, and villagers had long suspected that the well water was toxic.

Cancer Deaths in the Two Villages

We obtained a copy of a cancer death list compiled by the Liangqiao Village Committee from the Secretary, Ms. He, and double-checked it with Ms. Guo. Ms. He also provided us with the Liangqiao 'Register of the Deceased Population', which included information about deaths in Liangqiao from 1982 to 2007. Deaths from cancer were only part of this list, which the village committee files with the local government. The register recorded in detail

2 Southern Metropolis, 'Guangdong Wengyuan Cancer Village Research: 250 People Died Due to Pollution'.

3 Southern Metropolis, 'Guangdong Wengyuan Cancer Village Research: 250 People Died Due to Pollution'.

the name, address, time of birth, time of death, cause of death, and location of death of every deceased person, including even unnamed children who died on the same day they were born. Table 7.1 is a list of people who died of cancer between 1982 and 2007 that we extracted from the register.

Table 7.1 Liangqiao cancer deaths, 1982-2007

Number	*Family name*	*Year of death*	*Age of death*	*Type of cancer*
1	Zhu	1982	61	Oesophageal cancer
2	Zhu	1985	53	Liver cancer
3	Liang	1985	50	Throat cancer
4	He	1986	19	Liver cancer
5	Chen	1986	55	Lung cancer
6	Song	1990	51	Throat cancer
7	Zhang	1996	50	Lung cancer
8	Ye	1997	55	Lung cancer
9	Zhu	1998	61	Lung cancer
10	He	1998	59	Lung cancer
11	Zhu	1998	53	Stomach cancer
12	Zhu	1999	70	Lung cancer
13	He	1999	47	Liver cancer
14	Zhang	1999	87	Lung cancer
15	Guo	1999	55	Lung cancer
16	He	2001	67	Lung cancer
17	Liu	2002	73	Liver cancer
18	Wu	2002	54	Lung cancer
19	Guo	2004	67	Liver cancer
20	Zhu	2004	34	Lung cancer
21	Liang	2004	62	Lung cancer
22	He	2004	77	Lung cancer
23	Hu	2004	30	Colorectal cancer
24	He	2004	44	Uterine cancer
25	He	2005	60	Eye cancer
26	Zhang	2005	60	Oesophageal cancer

From Table 7.1 we can see that from 1982 to 2007, a total of 26 people in Liangqiao died of cancer, of whom 23 died in the twenty years between 1986 and 2005.[4] Liangqiao has a total of registered population of 382 people, so the annual average cancer mortality rate of the entire village during 1986 and 2005 was around 301/100,000, or nearly three times the rural cancer

4 Making individual statistics for the cancer death situation between 1986 and 2005 is to make it easier to compare with Shangba Village in the following session.

mortality rate for China in 1991 and 2000 (see Table 7.2). We can also see that the youngest person to die of cancer was only 19 years old, the oldest was 87, and the main types of cancer were lung, liver, oesophageal, stomach and throat cancer, which accounted for 50%, 15.4%, 7.7%, 7.7%, and 7.7% of the total. Lung cancer deaths accounted for half of all cancer deaths.

Table 7.2 Cancers mortality rates in 1991 and 2000 in certain regions of China (1/100,000)

Cancer type	*Urban*			*Rural*		
	1991	*2000*	*Change (%)*	*1991*	*2000*	*Change (%)*
Overall	123.92	146.61	18.31	101.39	112.57	11.03
Nasopharyngeal Cancer	2.17	2.14	-1.38	1.70	1.78	4.71
Oesophageal Cancer	8.94	9.06	1.34	16.32	15.24	-6.62
Stomach Cancer	19.69	19.94	1.27	22.55	20.94	-7.14
Colorectal Cancer	8.53	11.19	31.18	5.14	6.04	17.51
Liver Cancer	19.63	22.19	13.04	22.25	26.06	17.12
Lung Cancer	32.53	42.08	29.38	14.29	21.11	47.73
Breast Cancer	3.11	4.32	38.91	1.51	2.11	39.73
Leukaemia	3.60	3.82	6.11	3.31	3.18	-3.93
Cervical Cancer	1.49	1.08	-27.52	2.31	1.91	-17.32

Data source: Chinese Cancer Control Strategy Research Report, Bulletin of Chinese Cancer, 2002 Issue 5

As mentioned above, the town government helped us to contact the Shangba Village Committee and when we arrived in the village, the committee people came to receive us. However, they refused to provide a detailed list of deceased cancer patients. The conversation went as follows.

Q: We heard that you have statistical data on the patients?
A: None of the statistical data is public, the negative impact is too large. We still need to live in this place and think of future generations.
Q: We are conducting research. It is not for publicity, and it will not be used for anything else.
A: We still will not disclose it. After TV reports in 2005, there has been a negative impact. When we are selling vegetables, other people do not want them. It is a dishonourable thing.
Q: What is the general situation with the disease? Which year do the statistics end?
A: It is mainly liver, lung, stomach, and oesophageal cancer, and the statistics are from 1987 to 2004.

Q: The media reported 250 people contracting cancer within seventeen or eighteen years. Is that true?
A: It is basically true, there are that many people.
Q: How do you collect the data?
A: Hospital test results, hospital death certificates, and deregistration of deceased individuals.

Shangba was a famous 'cancer village' and had been reported on by most big media outlets, which had given relatively detailed descriptions of the cancer death situation. The research team compared and summarized reports from different media, and compiled statistics from the results (see Table 7.3).

Table 7.3 Main media reports concerning cancer deaths in Shangba

Media	*Report time*	*Cancer death situation*
CCTV Economy Half-Hour	January 8, 2001	More than 40 people died between 1997 and 2000, over 90% of them from cancer; of the 250 people who have died since 1986, 210 died of cancer, or 84% of all deaths.
CCTV Economy Half-Hour	June 3, 2005	The total number of Shangba Village cancer deaths has already reached 214 people.
Southern Metropolis Daily	Nov 18, 2005	In the eighteen years between 1987 and 2005, more than 250 people died of cancer.

The cancer death statistics collected by major media outlets were different due to the different time periods. Through interviews with the village committee and ordinary villagers we learnt that in the twenty years between 1986 and 2005, 214 people in Shangba died of cancer. Given that the total population of Shangba at the time was 3,400, the annual average cancer mortality rate of the village was 315/100,000. This is higher than the mortality rate in Liangqiao, which was 301/100,000, and more than three times the rural average cancer mortality rate for China in 1991.

Environmental Resistance Strategy

Active 'Problematization' in Liangqiao

Liangqiao is situated on the uppermost reaches of the Heng, and its paddy fields and farmlands were the most heavily polluted by the mine. Villagers believed that the main cause of the high incidence of cancer in the village was mining pollution. Before research institutes investigated the situation, villagers already had an awareness of the relationship between pollution and cancer through their ordinary knowledge and daily life experiences. They believed that their water sources, soil, and crops were contaminated, and that this was harming their health. They also adopted some coping strategies, for example using mountain spring water to replace well water as drinking water. Villagers believed that the pollution caused by mining not only damaged their health, but also contributed to their persistent poverty. They were unable to sell their crops, and even had to pay to get water from the mountains.

From Table 7.1 we saw that lung cancer accounted for half of all the cancer deaths in Liangqiao. We also found that smoking was very common among the villagers. However, when we mentioned whether the high incidence of lung cancer could be related to villagers' smoking, Ms. Guo countered, 'I do not think it is smoking, do people in other places not smoke?' Faced with a situation in which mining pollution might be causing a high incidence of cancer, Liangqiao villagers were not only forced to take concrete measures to deal with it, they also adopted strategies to 'problematize' (*wentihua*) the issue in the hope that Liangqiao would attract attention from the outside world, which would in turn generate government concern and lead to an improvement or solution to the problems of pollution and disease.

One practical measure the villagers took was to divert mountain spring water to replace well water for drinking. Liangqiao is on the upper reaches of the Heng and terrain is high. There are no other unpolluted rivers that can replace the Heng as a water source and so when the Heng became polluted and the water was no longer suitable for drinking, villagers joined together and each household contributed more than 1,000 yuan to install two- to three-inch plastic piping to bring spring water down from mountains. However, the amount of mountain spring water was always limited, and it could only solve the drinking water problem, not the need for water to irrigate the fields. At Ms. Guo's home, we saw that the irrigation ditch near the paddy field behind her house was full of polluted red-brown water. She said that villagers still used water from the Heng to irrigate fields, even though they knew doing so was harmful. They had no choice.

The population of Liangqiao was small, its economy was weak, and the power of the villagers was minimal. Villagers hoped to attract the government's attention through petitioning. But the Party Secretary, Ms. He, told us that the village committee had only four officials and in addition to the small number of staff, they also lacked funds. They did not even have the money to go petitioning. Because of this, Liangqiao cadres had never petitioned by themselves. They usually joined with other villages to petition various levels of government and environmental protection bureaus.

Ms. He discussed two petitions that had been organized in recent years. In 2005, along with other villages, Liangqiao jointly petitioned the County Environmental Protection Bureau (EPB), demanding that B mine compensate losses for 530 *mu* of polluted rice fields and 80 *mu* of polluted fish ponds in the village. Under the coordination of the EPB, B mine replied that even though it was partly responsible for the mining pollution in Liangqiao, other private mines also bore responsibility, and if the responsibility was not clearly shared, B mine could not give compensation. The Liangqiao Committee believed that pollution in Liangqiao originated primarily from B mine, and that it should bear the liability. When B mine did not agree to pay compensation, Liangqiao filed an appeal for an administrative review with the municipal EPB. In the end, with coordination of the municipal EPB, B mine agreed that 20 *mu* of paddy fields in Liangqiao were polluted and gave the village 168,000 in compensation.

In 2008, cadres from Liangqiao and more than ten other villages downstream that had been polluted went together to the Guangdong Provincial People's Congress and the provincial government to petition. After their petition materials were accepted, they were transferred to the Municipal People's Congress and the municipal government, and finally on to the county. After these persistent efforts by the villagers, B mine agreed to provide a total of 380,000 yuan in compensation. This money was allocated to each village by the county government, but Liangqiao Party Secretary Ms. He considered that the division of the money was also not fair.

In addition to petitioning, Liangqiao also actively sought help from the media and from experts and scholars. While researching in Liangqiao, our team could feel clearly the difference in attitude between the Liangqiao Committee and the Shangba Committee. The Shangba Committee needed a government notice to receive us, was unwilling to overtly disclose the situation regarding cancer, and even more unwilling to provide detailed information about cancer. In Liangqiao, the village committee received us enthusiastically, and provided us with all kinds of information and data. Ms. Guo from the village committee even personally accompanied us to see the waste water treatment plant and led us to her home to see materials relating to the petition and the lawsuit. She not only provided us with the complete registry of cancer deaths, but also voluntarily provided other researchers' articles and reports.

Why was the Liangqiao Village Committee so enthusiastic? Our team believes this has to do with their problematization strategy. The committee and the villagers hoped that the high incidence of cancer in Liangqiao could attract as much attention from the outside world as it had in Shangba, and that it could become equally 'well-known.' They hoped that outside influence might lead to a quicker solution of their problems, or at least to partial improvement. Committee cadres believed that Liangqiao was the primary victim of B mine's pollution, but the government did not care about their illness and suffering as much as they did about Shangba. The government did not solve the village's drinking water problem, and the amount of compensation the village received was also less than was given to Shangba Village. After seeing the 'accomplishments' Shangba achieved by problematizing the issue of cancer and pollution, Liangqiao villagers and leaders were convinced that this was the only way for the village to get outside attention and push the government to solve their problems.

Shangba switches to 'de-stigmatization'

In the 1970s, when Shangba villagers realized that the river water was being polluted, they started to look for substitute water sources, including ground water and mountain spring water. In the beginning, they dug wells in their own yards and switched to drinking ground water. The ground water in Shangba is abundant, and the villagers' wells were relatively shallow. But very soon, villagers felt that the well water also smelled, and was probably polluted as well. Some of them then thought of ways to divert mountain spring water and water from other unpolluted rivers. Near Shangba, there is a river in a village on relatively high ground that was still unpolluted. At the time of our research, around 20 to 30 households in Shangba were using water diverted from this river as drinking water. Every household contributed around 300 yuan, but the volume of water was quite small. Shangba Village has a total of 21 villager groups, and a population of more than 3,400 people. Only a minority was able to use mountain spring water and the rest had no choice but to drink their own well water. Some more creative villagers found ways to make money out of this situation. They drove their cars into the mountains to bring back spring water, and then sold it to other villagers. When the research team was chatting with a group leader in Shangba, the villager selling water happened to stop by his home. He said they had to go tens of kilometres from the village to collect the water. A big barrel was 30 kilograms, and sold for two yuan, and the water quality was very good. There were also a few villagers who rode motorbikes themselves to the mountains to bring back spring water for their families.

While doing whatever they could to avoid harm, Shangba villagers also actively organized petitions, partly aiming to solve the drinking water problem and partly to seek financial compensation. Shangba petitions were highly organized. The Village Committee used the cancer patient list as the basis for successive petitions, and villagers provided financial support for the Committee's efforts. As early as 1979, the village secretary went straight to Beijing to report the situation. In the 30 years since 1982, the 21 village groups have been responsible for raising money to support Shangba petitions. According to villagers, the Village Committee leases community land to others to plant trees, and they collect some rent from this that provided support for their many petitions.

Along with petitions, the Shangba Village Committee made full use of the media. CCTV and some foreign media reported on mining pollution in Shangba, and some research institutes in Guangdong came repeatedly to the village to conduct research. The unrelenting petitions from villagers and the widespread attention from the media put a lot of pressure on the local government. In 2008, the county government, municipal government, and B mine provided more than 14 million yuan in funds to build a reservoir and connect tap water to the homes of Shangba villagers. They also received financial compensation from B mine.

Ironically, while we were researching in Shangba, the village was in the process of repairing the drinking water infrastructure. Villagers said that the water from the reservoir was dirty and unfit for drinking, and that there were often water cuts. Villagers claimed that this was because the government misappropriated some of the project funds, cut corners in building the reservoir and left the water pipes exposed. Problems with the reservoir as a solution to the drinking water problem became the villagers' main concern, triggering a new round of petitioning. At the time of our research, the municipal government had allocated one million yuan for a maintenance fund for the facility, to be used for covering the exposed pipes.

Awareness of the negative impacts of 'stigma'

In contrast to the enthusiastic attitude of Liangqiao Village Committee cadres, when we arrived as agreed at the Shangba Village Committee, the cadres received us, but they avoided overtly discussing the topic of pollution and cancer. Typically, in areas with severe pollution and a high incidence of cancer, village cadres and residents usually attribute the high incidence of disease to pollution. However, the Shangba Village Committee cadres said they believed that 'right now nobody told us what the cause is, so we do not dare to say it's caused by pollution. We don't know what caused the cancer.' At first, we were very surprised by this attitude but certain things they said later explained it. Shangba Village had been fighting against pollution and the high incidence of cancer for a long time and in recent years their resistance had finally achieved some effect. The drinking water problem had received a preliminary solution, and some large companies also donated money to Shangba. Against this backdrop, village cadres started to consider the negative impact brought to Shangba Village by problematization, and this led them to consciously switch their approach to that of 'de-stigmatization'.

After some reflection, felt that this change in attitude was logical and in line with the actual situation. We had never encountered this situation before because the environmental resistance of 'cancer villages' had been in the early stages and had mostly not accomplished much. After a lot of hardship, the problem of pollution and cancer in Shangba had reached a certain level of resolution. The local government may also have exerted pressure on village cadres in the hope of avoiding a return to further petitioning. On the one hand, village cadres may have felt pressure from the local government, but at the same time they were also experiencing some of the negative impacts that problematization had brought to the village, including the difficulty young people had in finding marriage partners.

Comparison and Analysis

The difference in stages

Shangba and Liangqiao belong to the same township and are situated downstream and upstream of the same river. The relationship between pollution and cancer in the two villages also had many similarities. But the coping strategies the villagers adopted when faced with pollution and disease were very different. This reflects the fact that the two villages were at different stages in the process of environmental resistance. Liangqiao was still at the early stages and needed to adopt all kinds of problematization strategies to gain the attention of government and the outside world and push for a solution to the problem. Shangba also went through a phase of problematization but when its problem received a certain level of resolution, the villagers turned their attention to the negative impacts brought about by stigmatization, as well as the pressure exerted by higher levels of governments. This was why Shangba initiated efforts to remove the stigma.

Shangba and Liangqiao faced pollution hazards from the same source, and from the perspective of both existing medical research and common sense, the cause of the heavy metal pollution in the area was relatively clear. Judging from experience with mining pollution in other countries and regions, the causal relationship between heavy metal pollution and disease is also clear. But the awareness of the two villages towards pollution and disease, and especially their coping strategies, were different, and this led their problem-solving strategies to be drastically different. Shangba was slightly 'overly stigmatized', and so they were beginning to try to 'remove the stigma' and cared less about the relationship between cancer and mining pollution. Liangqiao was closer to the pollution source and more severely affected by pollution, but it was neglected by the government and unknown to the outside world. Because life was hard in Liangqiao and the villagers wanted to attract attention to their plight, they were strongly motivated to attribute cancer to mining pollution, and sought to be recognized as a 'cancer village'.

Economic Factors

Economic factors were also important in causing Shangba and Liangqiao to choose different resistance strategies that brought different results. The two villages are both located in rural northern Guangdong, and their economic circumstances were not good. Comparatively, Liangqiao was in a worse situation. The main source of agricultural income for both villages was sugar cane. Shangba villagers said that during the season, the market price for sugar cane could be as high as 1,000 yuan/ton, but as with most agricultural products, it was mostly the middle men, not the farmers, who made money from its sale. Mr. He, the leader of the village group near the Shangba Village Committee office, described his situation. His family rented three *mu* of land to plant sugar cane that year. The gross income was over 20,000 yuan, but the costs were 10,000 yuan, including expenses for fertilizer and rent, so the profit his family made from sugar cane that year was only around 10,000 yuan. In recent years, because the village had become known as a 'cancer village', rice did not sell well. Villagers reduced the area planted with rice to just enough to satisfy their own needs. Even Shangba, with its limited income, was still a rather prosperous village and its residents were rich in the eyes of people from Liangqiao. Liangqiao Village Committee cadres believed that the local government paid more attention to the economically stronger Shangba and neglected the poorer Liangqiao.

The influence of economic strength on villagers' approach to environmental resistance was mainly reflected in their ability to sustain funding for petitioning in terms of travel costs and expenses for materials. In Shangba, funds were first collected from each villager group, and then pooled by the village committee to support petitions. Spread across 21 villager groups and 3,400 people, 10,000 yuan in petitioning fees over 30 years is actually a very small sum – just a few yuan per person. But the entire village of Liangqiao had only four groups and just over 380 people, so even if the petition fees were shared it would not be a small expense for villagers.

Differences in clan power

In the environmental resistance of Shangba, aside from the population and economic strength, clan power also had an important effect on the solution of the pollution problem. Even though Shangba is a relatively large administrative village, the main family name is He, and residents originate from the same ancestral clan. The administrative village, kinship, and geographical relationships basically overlap with each other. In the period prior to our research, rural clan activities had been reviving. While our team was investigating in Shangba Village, we happened to observe the ceremony marking the renovation of the Shangba ancestral hall. The whole ceremony was very grand. Many male clan members had red flowers pinned to their chests, and we were told they had donated to the renovation fund. In the eyes of Liangqiao, Shangba villagers had strength in numbers, and what they said carried weight. In fact, the size of a village is not directly related to its strength, and the reason why Shangba was powerful may be more related to the power of its ancestral clan.

Typically, clan organizations are social groups with strong internal recognition and clear external boundaries, but the influence of the Shangba clan was somewhat concealed. This was partly because Shangba was an administrative village, and this diluted the presence of its clan organization, but also, to the extent that villages are self-governing, the clan could make use of its members' participation in these self-governance institutions to manage local affairs, making the its role appear more 'legitimate' and 'official'.[5] In villages with relatively strong clans, their power naturally influences local village elections, and through this affects the selection of officials to the township or even higher levels of government. We noticed that in the township government, quite a few officials came from Shangba. Liangqiao Village Committee cadres also mentioned that the power of Shangba not only influenced their township, but also even higher levels of government.

5 Du, *Culture, Power, and State*, p. 80.

The Dilemma of the Local Government

At the stage of development these villages were in when we conducted our research, local government often faced the dilemma of making a choice between economic development and environmental protection. Local government officials needed to maintain the pace of economic growth, but they also faced the pressure of environmental pollution, and especially pressure of continuous petitioning by villagers who were the victims of pollution. B mine is a large state-owned enterprise, and its financial situation directly affected local government revenue. When considering the relationship between economic development and the environment, maintaining local economic growth was incredibly important. In addition to that, B mine was a provincial level state owned enterprise – a relatively high administrative level – and the county level government did not have much power to interfere with it. This made environmental regulation and control more difficult.

Because Shangba was a famous 'cancer village'. There were often media reporters there, and even foreign journalists, as well as all kinds of research studies going on. The local government also faced a dilemma. On the one hand, they hoped the problem could be resolved, pollution could be reduced, disease could go down, and everything would be fine. But they also had other concerns. In the end, pollution and the cancer problem were both negative kinds of news and reporting them to the outside world would be damaging to the local image and a hindrance to future development. In the view of some villagers, the local government was an obstacle in the process of solving environmental problems. For example, if Liangqiao cadres gave interviews to foreign media correspondents, afterwards they often faced pressure from government officials. Legal experts, professors, and students from a university in Beijing visited Liangqiao multiple times, wanting to help villagers with a lawsuit or health check-ups, but after a few years, things progressed slowly. Liangqiao cadres were fully aware of the complicated relationship between the local government and the polluting enterprise, and the many challenges this entailed.

In short, the case studies of Shangba and Liangqiao produced some findings that differ from studies of other 'cancer villages'. First, we found that in Shangba and Liangqiao, the extent to which the pollution problem had been solved and the satisfaction of the villagers were inversely related to the way in which villagers considered the relationship between pollution and cancer. In Liangqiao, we found that the villagers pointed to pollution from mining as the cause of cancer, but this was not because that they had scientific knowledge or evidence, but because doing so was the only way they could make strategic use of the problem and get a solution. For ordinary farmers, the negative impact of pollution and cancer was heavy, and this made them eager to solve the problem; but the actual cause of cancer was not really their concern. Suffering prevailed over the truth of the matter.

Second, Shangba and Liangqiao adopted very different strategies because they were at different stages in their environmental struggle. In previous investigations, we had the impression that all villagers in 'cancer villages' were victims and that they would be willingly tell outsiders of their suffering and pain. The attitude of the Shangba Village Committee changed our expectations. For Shangba, 'stigmatization' had initially brought benefits in terms of solutions for the problems of pollution and disease. They had their drinking water problem solved and also received economic compensation. By this time, they had started to care about the negative influence of 'stigmatization' on the village and were attempting to engage in 'de-stigmatization'. Lastly, we have noticed that clan power continues to play an important role in managing rural public affairs. This and other social factors are new issues that deserve further discussion and research.

Works Cited

Du, Zanqi. *Culture, Power, and State*. Translated by Wang Fuming. Nanjing: Jiangsu People's Publishing House, 2010.

Southern Metropolis Daily. 'Guangdong Wengyuan Cancer Village Research: 250 People Died Due to Pollution', 18 November 2005.

8 Behind the 'High Incidence of Lung Cancer'[1]

Li Qi and Chen Ajiang

Abstract

After 2008, a movement opposing waste incinerators swept across many cities in China. At the time, there was already a conflictual relationship between L waste incinerator in Guangdong and surrounding residents, but the 'Panyu incident' caused what was originally a small local conflict to become an issue of wider and even national concern. The list of 'cancer victims' that circulated on the internet, and related media reports, gradually convinced the public that the waste incinerator plant was responsible for high rates of lung cancer among residents. However, our investigations found that there were numerous errors in the information about the incidence of lung cancer. The chapter discusses what we discovered about the complex background to this case, and the reasons why rumours and misinformation spread so easily at this time.

Keywords: cancer villages, rural industrial pollution, environmental conflict, internet rumours, social structure

Along with rapid economic development, urbanization and improving living standards, the amount of municipal household waste in China has been steadily increasing. Although investment in waste treatment facilities has also been rising, it is still hard to keep up with the demand. Because landfills require quite a lot of land, and can easily create secondary pollution, finding sites for them is becoming increasingly difficult. Incineration effectively reduces waste, takes up less space, and has clear technological and economic advantages. However,

1 This article was published under the title 'On "Fake" Data of High Incidence of Cancer, Based on Research of Sociology' in the *Journal of Nanjing University of Technology* (*Social Science Edition*), Issue 1, 2013.

Chen Ajiang, Cheng Pengli, and Luo Yajuan, *Chinese "Cancer Villages": Rural Development, Environmental Change and Public Health*. Taylor & Francis Group 2020
DOI: 10.5117/9789089647221_CH08

in recent years, there have been continuous mass incidents opposing waste incineration power plants in China. In 2009, in particular, Beijing, Shanghai, Guangzhou, Nanjing, Wujiang, and other areas all saw 'anti-incineration' movements. Strong public opposition caused some waste incineration power plant projects (we will abbreviate to 'incinerators' hereafter) to be stopped or postponed during the feasibility, planning or construction stage. Of these, the 'Panyu Incident' in Guangzhou had the biggest social impact. Continuous coverage of the incident in the media over several months, and fierce arguments and relentless debate between 'pro-incineration' and 'anti-incineration' groups suddenly pushed the issue of waste incineration to the centre of public debate.

The wide impact of the 'Panyu Incident' attracted our attention and in July 2010, Li Qi headed to Guangzhou to research the situation. During the investigation, she unexpectedly discovered that L waste incineration power plant (hereafter referred to as L Incinerator) was said to be causing severe pollution, causing neighbouring residents to have a high incidence of cancer. Later, she collected a lot of information about the plant, and she also accessed the internal online residents' discussion forum relating to the 'Panyu Incident' in order to understand the situation better. In December 2010, the research group went to the area near L Incinerator to conduct fieldwork. We communicated with villagers and village officials in communities near the plant and conducted interviews at the facility itself and with the Environmental Protection Bureau (EPB), the Centre for Disease Control (CDC), and other organizations. In the process of sorting through and analysing the data from our fieldwork, we discovered that local claims of a high incidence of respiratory cancer occurring after L Incinerator was built were at odds with the findings of medical science that there is a considerable time lag between the occurrence of pollution and the onset of cancer. This prompted the second phase of our investigation. In June 2011, Li Qi conducted home interviews with the families of people on the 'list of cancer victims' and collected a lot of new information.

In developed countries, environmental pollution occurred relatively early, and conflicts and contradictions related to pollution also got a lot of exposure, triggering a wave of research on environmental justice, environmental resistance, and related topics. Over the last decade, the problem of environmental pollution has become very pronounced in China, causing frequent disagreements and conflicts. There have been a lot of studies of the issue, for example, research by Luo Yajuan about villagers in Dongjing in Northern Jiangsu being affected by pollution and engaging in resistance,[2] or Jing Jun's social and cultural analysis of environmental pollution and

2 Please see Chapter 3 of this book.

villagers' interpretations of it in Dachuan Village in Gansu,[3] and Feng Shizheng's examination of environmental resistance from the perspective of the basic structure of relationships in Chinese society.[4] The term 'environmental resistance' as it is used in environmental sociology research in China focuses mainly on examining the way in which 'the people' resist 'enterprises' or 'the government', and implies the hypothesis that protests are undertaken by 'vulnerable groups' against 'the powerful'. In terms of past experience, this hypothesis basically seems to conform to the actual situation. However, in reality, even 'the weak' have 'the weapons of the weak' that they can deploy in their dealings with enterprises and local government. Or in other words, there are situations that do not conform to this hypothesis.

For example, as we discovered in our 'cancer village' research, some residents make use of the word 'cancer' as a way to gain economic benefits or other forms of social support.[5] Because waste disposal sites pose health and financial risks to the people who live around these facilities, siting landfills is very difficult. Foreign scholars typically use the term 'NIMBY' (Not In My Back Yard) to describe this phenomenon. There has already been a lot research on the 'NIMBY' phenomenon in the process of siting waste dumps in China, including some case studies. He Yanling attempted to analyse the special characteristics of 'Chinese style' NIMBYism through an incident of residents' protest against an electricity transformer substation.[6] In terms of research on health hazards and environmental pollution, a classic example is Minamata Disease in Japan.[7] Martha Balshem's research about the 'cancer community' in Philadelphia also demonstrates the different perceptions among local residents, scientific authorities, and doctors.[8] The situation in the case we discuss here contains features similar to those in the research mentioned above, but also characteristics that are specific to this particular place and time. This chapter focuses on these.

The Disputed 'High Incidence of Cancer'

Information on the internet and in the media reported a high incidence of cancer among people living around L Incinerator, but information released

3 Jing, 'Consciousness and Awareness: Environmental Resistance of a Xibei Village', pp. 5-14.
4 Feng, 'The Silent Majority: Chaxu Geju and Environmental Action in China', pp. 122-132.
5 Please see Chapter 9 of this book.
6 He, 'Chinese Style' NIMBY Conflict: Analysis Based on Events', pp. 102-114.
7 Funabashi, 'Minamata Disease and Environmental Governance', pp. 7-25.
8 Balshem, *Cancer in the Community*, p. 59-71.

by the CDC stated that the incidence of cancer was not high. Despite this discrepancy, the huge volume of information circulating on the case and the attitude of those participating in the discussion showed that the 'high incidence of cancer' had already been commonly accepted by the public.

'List of Cancer Victims' on the Internet

The 'Panyu Incident' was a public protest against the construction of an incinerator in Panyu in Guangzhou. On 23 November 2009, the conflict escalated, and nearly a thousand citizens gathered in front of the city government. Purporting to be 'taking a stroll,' they demanded that the government halt construction of the incinerator. The public carried 'anti-waste incineration' banners, and some residents held slogans saying, 'Y villager already has terminal cancer'. Previously, on 18 October 2009, residents of Panyu had also conducted an investigation in the area around L Incinerator. They interviewed local residents and made a video titled 'What can we do to Save the People of L?', which they uploaded onto the internet. On 23 November, the 'mass stroll' attracted media attention, a lot of information about the event was circulated, and this made the issue more complex. The 'strong evidence' behind the accusation that pollution from L Incinerator 'was causing cancer' was a copy of a list of cancer victims (see Tables 8.1, 8.2 and 8.3). As soon as it was released the name list caused an uproar, and it was enthusiastically forwarded by netizens on all the major online forums.

Table 8.1 List of cancer deaths in Y Village before the plant was built (1993-2005)

Number	*Community*	*Name*	*Gender*	*Date of Birth*	*Date of Death*	*Age*	*Cancer Type*
1	2	Zhou*Zuo	Male		2004.5.14		Nasopharyngeal cancer
2	4	Zhou*Liang	Male	1953.10.6	2002.11.16	49	Liver cancer
3	5	Xu*Bo	Male	1954.4.16	2004.11.14	50	Liver cancer
4	12	Zhang*Ying	Female		2002		Breast cancer
5	12	Fan*Xin	Male		1998.6		Liver cancer
6	14	Zhou*Qiang	Male	1928	1996	68	Lung cancer
7	16	Zhou*Hua	Male		2004.11.28		Leukaemia
8	17	Zhou*Cai	Male	1952.12.14	2004.9.6	52	Stomach cancer
9	18	Chen*Xia	Female	1961.6.19	2003.6.5	42	Breast cancer

Original note: The above data was provided by Y villagers and is incomplete. It also does not include permanent residents without Y residence registration.
Data source: http://gzbbs.soufun.com/2811072106~-1~487/55665120_55665120.htm, Accessed September 10, 2010
Author's note: '*'is used hereafter to protect the privacy of the parties concerned.

Table 8.2 List of cancer deaths after the plant was built (2005-2009) in Y Village

Number	*Community*	*Name*	*Gender*	*Date of birth*	*Date of death*	*Age*	*Type of cancer*
1	1	Su*Tu	Female	1952.12.01	2008.03.08	56	Nasopharyngeal cancer
2	2	Zhou*Zhen	Male	1953.08.27	2008.07.10	55	Liver cancer
3	3	Zhou*Heng	Male	1955.06.08	2006.05.21	51	Lung cancer
4	4	Zhou*liang	Male	1945.02.12	2006.10.10	61	Liver cancer
5	5	Xu*Gui	Male	1941.10.05	2008.08.28	67	Lung cancer
6	5	Xu*Rong	Male	1933.12.12	2008.12.09	75	Lung cancer
7	6	Chen*Gui	Male	1959.03.02	2008.01.14	49	Nasopharyngeal cancer
8	8	Zhang*Rong	Male	1922.09.17	2007.09.30	85	Lung cancer
9	8	Chen*Feng	Female	1944.03.19	2008.08.18	64	Throat cancer
10	8	Chen*Qing	Male	1939.12.02	2009.10.12	70	Lung cancer
11	9	Liu*Huang	Male	1951.08.08	2007.01.14	56	Nasopharyngeal cancer
12	9	Chen*Ying	Female	1952.05.21	2008.01.15	56	Lung cancer
13	9	Chen*Sheng	Male	1945.04.08	2008.04.28	63	Lung cancer
14	10	Fan*Rong	Male	1961.12.18	2009.06.07	48	Lung cancer
15	10	Fan*Bin	Male	1947.07.19	2008.01.23	61	Lung cancer
16	11	Xu*Ping	Female	1960.05.10	2009.09.28	49	Lung cancer
17	11	Fan*Xin	Male	1936.06.27	2008.02.25	72	Lung cancer
18	11	Fan*Qin	Male	1937.11.20	2007.04.01	70	Lung cancer
19	12	Fan*Fa	Male	1949.07.20	2007.03.15	58	Lung cancer
20	12	Fan*Biao	Male	1960.02.17	2007.12.20	47	Lung cancer
21	12	Fan*Cai	Male	1939.11.01	2009.09.20	70	Lung cancer
22	13	Zhuang*Sun	Male	1954.10.23	2009.03.13	55	Nasopharyngeal cancer
23	13	Chen*Hao	Female	1956.11.05	2009.08.28	53	Lung cancer
24	13	Chen*Yuan	Male	1942.12.05	2009.10.13	67	Lung cancer
25	13	Xu*Ying	Female	1956.11.05	2009.08.07	53	Lung cancer
26	14	Zhou*Xiang	Male	1940.08.14	2006.01.23	66	Nasopharyngeal cancer
27	14	Zhou*Ming	Male	1959.10.03	2006.08.25	47	Liver cancer
28	14	Ye*Wei	Female	1943.09.01	2008.06.24	65	Lung cancer
29	14	Zhou*Qian	Male	1934.05.05	2008.07.26	74	Lung cancer
30	14	Zhou*Ling	Male	1939.11.01	2008.12.10	69	Lung cancer
31	14	Zhou*Lang	Male	1950.03.11	2009.04.22	59	Lung cancer
32	14	Zhou*Tong	Male	1971.09.09	2008.01.05	37	Lung cancer
33	14	Zhou*Cang	Male	1937	2006.11	69	Lung cancer
34	14	Zhou*Cang	Male	1940	2006	66	Lung cancer
35	14	Cao*Qing	Female		2007		Liver cancer
36	16	Zhou*Jiu	Male	1949	2007.09.28	58	Lung cancer
37	17	Zhou*Ming	Male	1936.05.08	2008.08.27	72	Lung cancer

Number	Community	Name	Gender	Date of birth	Date of death	Age	Type of cancer
38	18	Zhang*Di	Female	1934.03.06	2009.07.20	75	Stomach cancer
39	18	Zhou*Qing	Male	1962.09.29	2006.03.08	43	Nasopharyngeal cancer
40	18	Zhou*Ming	Male	1972.12.07	2007.08.14	35	Liver cancer
41	19	Xu*Ling	Male	1966.01.22	2006.05.27	40	Liver cancer
42	19	Xie*Lian	Female	1936.05.10	2007.07.11	71	Uterine Cancer

Note: The above data was provided by Y villagers and is incomplete. It does not include permanent residents without Y residence registration. The 'Fan*Biao' in this list is the spouse of 'Huang*Tian' in the following list of 'Surviving cancer victims in Y Village'. 'Zhou Yu*' (cancer) and 'Zhou Zhen*' (lung cancer) in Community 14 are brothers. Because the type of cancer Zhou Yu had and the exact date of death for both of them still need to be confirmed, they are temporarily not included in the lists. Other, similar situations include 'Zhou*Rong' of community 18 and 'Zhou*Quan* of Community 14. These four patients have not been counted in later data analysis. Data source:http://gzbbs.soufun.com/2811072106~-1~487/55665120_55665120.htm, Accessed September 10, 2010

Table 8.3 Surviving cancer victims in Y Village after the plant was built

Number	Community	Name	Gender	Type of cancer
1	1	Xie*Qiong	Female	Colorectal cancer
2	2	Geng*Huan	Female	Breast cancer
3	2	Zhou*Wen	Male	Nasopharyngeal cancer
4	3	Zhou*Quan	Male	Colorectal cancer
5	7	Chen*Chi	Male	Nasopharyngeal cancer
6	9	Chen*	Male	Colorectal cancer
7	9	Chen*Ming	Male	Nasopharyngeal cancer
8	10	Chen*Di	Female	Nasopharyngeal cancer
9	11	Fan*Hua	Male	Lung cancer
10	12	Huang*Tian	Female	Breast cancer
11	12	Fang*Zi	Male	Lung cancer
12	12	Fan*Sheng	Male	Lung cancer
13	13	Shen*Dai	Female	Uterine cancer
14	13	Chen*Yuan	Male	Nasopharyngeal cancer
15	14	Zhou*Chang	Male	Lung cancer
16	15	Pan*Yun	Female	Uterine cancer
17	18	Lin*Di	Female	Lune cancer
18	18	Fan*Huan	Female	Colorectal cancer
19	18	Chen*Di	Female	Lung cancer
20	19	Xu*Xia	Female	Uterine cancer

Note: The above data was provided by Y villager and is incomplete. It also does not include permanent residents without residence registration.
Data source: http://gzbbs.soufun.com/2811072106~-1~487/55665120_55665120.htm

From these lists, we can see that:

1 The number of cancer deaths greatly increased after L Incinerator was built compared with the period before. In the thirteen years (1993-2005) before the plant was built, there were nine cancer deaths in Y Village, or an annual average of 0.69 people. In the four years after the plant was built (2006-2009) there were 42 cancer deaths, for an annual average of 10.5 people. The annual average number of cancer deaths after the plant was built was about fifteen times that of the average before it was constructed. Additionally, there are four cases that were not counted and twenty surviving cancer victims.

2 Respiratory tract cancers have a high mortality rate. Among the nine cancer deaths before the plant was built (1993-2005) there were only two cases of respiratory tract cancer (lung cancer and nasopharyngeal cancer), accounting for 22.2% percent of all cancer cases. Of the 42 cancer deaths after the plant was built (2006-2009) 34 cases were respiratory tract cancers, accounting for 80.95% of the total. The average annual number of deaths from this type of cancer after the plant was built was around 57 times the number before the plant was built. Furthermore, there were twenty cases of confirmed surviving cancer victims, of which eleven cases were respiratory tract cancers, accounting for 55%.

The impression given by the list of cancer victims on the internet is that after L Incinerator was built, the number of cancer cases among people living around it clearly rose, especially respiratory tract cancers such as lung cancer, which increased rapidly. This list 'with full names' convinced a lot of people that L Incinerator caused the high incidence of cancer among the villagers.

Dissemination and Interpretation by the Media

The name list on the internet also caught the attention of the media. On 26 November 2009, the CCTV 'News 1+1' programme, 'Confronting Garbage, Public Opinion is Gold' reported in detail the situation of the Guangzhou City Panyu Waste Incineration Power Plant project, which was facing strong opposition from residents. The programme also mentioned L Incinerator,

> According to Panyu residents' self-initiated investigation, people living around the L Incinerator, which was built four years ago in another part of Guangzhou City, have reported an increased number of cancer victims in recent years.[9]

9 'Facing Garbage: Public Opinion is Gold', CCTV News Channel 'News 1+1' Programme, 26 November 2009, http://tv.people.com.cn/GB/166419/10450967.html, Accessed September 15, 2010.

A report in 'China Newsweek' on 30 November 2009 entitled 'L, 'We are Sad for You' officially reported the cancer situation around L Incinerator.

> The three communities 10, 11, and 12[10] that are less than 400 metres away from L Incinerator have a population of 180, 350, and 320 people respectively. In the twelve years between 1993 and 2005, these three villages did not have a single case of respiratory tract cancer, and only Zhang Meiying of Community 12 died of breast cancer. However, over just three years since the power plant started operating in 2006, these three communities suddenly had thirteen cases of cancer. Community 14, which is less than 800 metres away from the power plant, has a population of 300 people. Between 2006 and 2009, ten people died of cancer; and with the exception of Cao Guiqing, who died of liver cancer, the others all died of two types of respiratory tract cancer – lung cancer and nasopharyngeal cancer.[11]

This article listed the names, community, and cancer information of twelve cancer victims (mostly the same as the information on the 'cancer victim list' on the internet). While investigating, the reporter also found that the black residue from the '979 degree' furnace contained debris from rope, fabric, red plastic bags, and even sports shoes which had not been completely incinerated. It was therefore suspected that the temperature of the furnace did not reach 850°C, meaning that it would produce a large amount of the level one carcinogen, dioxin.

Later, a 'Wealth Investment Weekly' report on 7 December 2009, entitled 'L High Incidence of Cancer and Its Correlation and Causality with Waste Incineration Projects' made a statistical analysis of the 'Y villager cancer victim list' online. It concluded,

> Judging by the logical chain formed by consistency between three variables— before and after the establishment of the factory, the length of time, and the distance – it can be confirmed that there is a high correlation between the power plant and the high rate of cancer and respiratory diseases in Y Village. We can even establish a causal relationship based on this correlation. The reason for the high incidence of cancer, and especially cancer of the respiratory system, is that the smoke discharged by the power plant contains dioxin, furans, and other organic pollutants

10 The 'community' here refers to economic partnership communities, equivalent to 'villager groups' in other rural areas and is used hereafter.

11 Chen, 'L, We are Sad for You'.

> such as lead, mercury, cadmium, and other heavy metals as well as dust (especially dust particles with a diameter less than 2.5 millimetres, which cannot be expelled once they have been inhaled). These types of flue gas are full of all kinds of carcinogens. They are inhaled day and night by nearby residents and build up inside the body over time. Under the combined and cumulative effects of these different carcinogens, it can be predicted that the incidence of cancer and respiratory tract cancer among nearby residents will continue to increase, and that the average age of death will continue to fall.[12]

This report used 'scientific' terminology to analyse the relationship between the L Incinerator and cancer. Quite a lot of people were convinced by this analysis and used it as 'authoritative evidence' to support their accusations against the plant.

CDC Investigation Results

In response to the spread of the cancer name list, the Municipal Centre for Disease Control (CDC) teamed up with the district CDC and on 3 December 2009 it conducted preliminary investigations on the cancer situation among Y Village residents. A second round of investigations was conducted from 12-22 January 2010. The CDC published the results of these investigations on 26 January 2010.

The CDC conducted verification of the 'list of Y Village cancer victims' that was trending on the internet. The results were: (1) in Table 8.1, nine people on the 1993-2005 Y Village cancer death list were confirmed to have died of cancer, but there were errors regarding the types of cancer from which three of those people had died: Zhou*Liang had pancreatic cancer, Chen*Xia had liver cancer, and Zhou*Qiang had colorectal cancer. Cause of death registration data at the police station also showed that a further 27 people in Y Village had also died of cancer between 1993-2005; (2) of the 42 cases of cancer victims in Table 8.2, 18 were eliminated, 22 were verified, and there were two (Zhou*Cang, Cao*Qing) that could not be verified because the entire family had relocated and they could not be tracked down; (3) the twenty cancer victims who were surviving after the plant was built (Table 8.3) were all Y villagers, and with the exception of Fan*Hua (who died on 26 December 2009), the others were all alive. The investigation

12 Qing, 'L High Incidence of Cancer and Its Correlation and Causality with Waste Incineration Project'.

eliminated three of these cases, and seventeen were verified; three of whom had contracted cancer before 2003.[13]

In addition to verifying the situation of the people on the internet cancer list, the CDC also conducted statistical analysis of the overall mortality rate, and the incidence of all cancers and of lung cancer among Y villagers between 2003 and 2008. The results show that there was no significant change over those years in the overall annual mortality rate, cancer incidence or lung cancer incidence among Y villagers. The conclusion of the CDC investigation was that: (1) 21 of the cases on the 1993-2009 list of Y Village cancer victims were eliminated through investigation, 48 were verified, and two were unconfirmed because they could not be contacted; (2) of the 72 total cases of cancer in Y Village between 2003 and 2008, 35 individuals contracted the disease between 2003 and 2005, and 37 between 2006 and 2008; the onset rate of cancer among Y villagers did not change significantly between 2003 and 2008; (3) between 2003 and 2008 there were thirteen cases of lung cancer in Y Village, in four of which the individual contracted cancer between 2003 and 2005 and in nine between 2006 and 2008; the onset rate of lung cancer did not change significantly between 2003 and 2008.[14]

The results of the CDC's investigation were extremely disappointing to many people who opposed waste incineration. Many citizens suspected that the CDC was biased to some extent and did not report the complete truth to the public. The results of the investigation were not trusted by the public and did little to dissipate their suspicions, doubts, and accusations. The social issue of 'cancer villages' was not resolved.

Research Findings

In December of 2010, our team went to L Incinerator, the neighbouring villages, and related government departments to conduct research. The situation we learned about through fieldwork was very different from what the internet and the media had understood to be the case. Further research was conducted during June of 2011, and the team discovered that the information about lung cancer on the 'cancer name list' was seriously inaccurate.

13 For more information, see the city CDC's investigation report on the Y Village cancer situation: *Guangzhou Daily*. 'Y Village cancer Victim Name List Rumor Inaccurate', p. A7.
14 Ibid.

The Overall Situation of the 'Cancer Village'

The place that was described on the internet as 'severely polluted, with a high incidence of cancer' turned out not to be a remote and shabby hamlet, but a large and relatively economically developed village, more like a thriving small town. One day in early winter 2010, our five-member team walked along a broad road towards the high chimneys of the incinerator plant. We did not feel anything particularly out of the ordinary, and it was not 'unbearably smelly', as netizens had previously described it. Occasionally garbage trucks passed by, but they were all sealed, and there was no noticeable stench. We arrived at the gate, and after we explained who we were, the guard helped us to contact the office, but the staff replied that an official letter was needed before they could receive visitors. For the time being we therefore observed the environment outside the plant. There were no strange odours in the area nearby, nor did we see any noticeable pollution in the stream that ran along the side of the facility. The chimney discharged a continuous plume of white steam[15] (see Photograph 20) and there was an online electronic monitoring screen at the door which broadcast real time monitoring data for sulphur dioxide, soot, dust, and other major pollutants produced during waste incineration (see Photograph 21). Our initial sense was that pollution from the incinerator was not as severe as it had been described, and the situation was very different from what we had expected.

Through visits with villagers and the village committee, we gradually learned more about the situation. First, we clarified the actual relationship between 'L Village' and 'Y Village'. It turned out that these had been conflated on the internet and that 'L Village' does not actually exist. The administrative village where L Incinerator is situated is 'Y Village'. The natural village nearest the plant is H Village, and it includes Groups 10, 11, and 12 of Y administrative village. 'L' was originally the name of a small reservoir in Y Village, and in 1989 it was chosen as the site of a waste landfill, which was named L Household Waste Landfill. Because the incinerator was later also sited near the landfill, it was also the called L Incinerator. The 'People of L', who are frequently mentioned on the internet, probably refers to villagers living around the incinerator, but there is no actual village or group with that name. Many outsiders mix these places up.

15 People on the internet reported that the incinerator's chimneys discharged thick black smoke. However, from our observation during fieldwork, the smoke rising from the chimney was water vapor produced by the furnace when burning coal or other fuels.

Y Village is part of T township, which is one of the fifteen core townships of the city. The central district of the town is 18 kilometres from the centre of Guangzhou city. The 21 administrative villages in the township have a total population of 120,500 people. Of these, 71,300 have local household registration, and around 50,000 are long-term residents from outside. Y Village is situated in the south of T Town. There are two main roads running through the village from east to west, providing convenient transport. The total area of the village is 8.8 square kilometres, with 2,567 *mu* of arable land. As of the end of 2010, Y Village had 7,390 residents with local household registration, and a migrant population of around 35,000 people. The village has a total of 23 natural villages, divided into 19 groups with registered populations of 222 to 544. Y Village has used its favourable geographical location and previously unexploited resources to develop a rental sector and attract investment, while also developing goods distribution and warehousing businesses, as well as residential housing, food service and other tertiary industries. These have injected new vitality into the development of the village. After 2000, the village committee developed unused land near the mountains and raised capital to construct factory buildings and attract outside enterprises. At the time of our research, the village had more than 200 enterprises and the local economy took off as the result of this real estate development. In 2009 villagers' annual average income reached 13,192 yuan, making it the village with the highest collective income in T town. The factory cluster completely changed the traditional rural look of Y Village; its basic infrastructure and living environment both improved and the village had a complete range of service facilities.[16]

At the time of our research, the factories in Y Village were mainly producing clothing, shoes, leather goods, food and handicrafts. There were also a few cosmetics companies. Local residents basically did not engage in agricultural production anymore; their income came primarily from share dividends, leasing out houses, and migrating for work. The majority of the village's original land had been rented out and built on, and the few remaining vegetable plots had also been contracted to outsiders. Rent from factories and vegetable plots was divided by the community among villagers according to the number of people in the household and their age. Many young people in the village were working elsewhere, and the elderly were raising their children for them at home. The villagers' houses were mainly three to six floors and renting space to migrant workers provided a stable income.

16 Information provided by Y Village Committee.

Villagers Report 'Severe Pollution, and a Lot of Cancer'

When we visited villagers, every time L Incinerator was mentioned, there would always be people who expressed their discontent, especially villagers in Groups 10, 11, and 12 which were closest to the facility. Their reaction was the strongest. A frequent claim was that dense black smoke from the plant's chimney caused a thick layer of dust in their homes. Some villagers said,

> It's not too bad during the day, but at night it stinks so much we dare not even open the window. It's because what they burn during the day and what they burn at night is different. At night they burn chemical products and plastics. If there's wind it's better, but when there's no wind it smells even worse. The grain (rice) that we used to plant looked really pretty as the water flooded the fields, but the seeds never sprouted. A lot of people in the village get cancer every year; the youngest are only thirty or forty years old. Before the plant was built, we never heard there were so many. (Ms Fan, Y Village Group, June 12, 2014).

Some of the villagers have learnt a little about 'dioxin' from TV or the internet, and have applied this information to the local situation:

> You won't see that kind of smoke during the day but at night they secretly discharge a lot of smoke. That kind of ash is very serious. The roofs of the villagers' houses around here all have a layer of black dust, and that is the most toxic; all the dioxins in there. (Mr Zhang, Y Village Group 8, June 12, 2011).

Villagers living along the road along which waste was transported reported that when the garbage trucks passed, they gave off a very strong odour. They said it was especially intense every morning around five to six o'clock and every afternoon around four to five, when a lot of garbage was being transported. Some villagers also reported that waste water from the plant was being discharged into the surrounding vegetable plots and fields, and that they did not dare to eat vegetables grown in this water. The plots in the area were mostly contracted to outsiders, who sold the vegetables to other places. Many people mentioned that the incinerator 'does not burn during the day, it burns at night', and they suspected that it was secretly discharging waste. But the issue that worried the villagers most was cancer, which they said had become more prevalent in recent years, and for which they blamed the incinerator.

Power Plant Claims 'Standard Operations'

After some coordination, our request to visit L Incinerator was granted, and our team visited the plant on 29 December 2010. The employees gave us a tour of the entire facility, explaining the technical process, the flow of waste through the treatment systems, and other aspects of the plant's operations. We learnt that this was the first waste incineration power plant in the city and that it was rated a 'National Key Environmental Protection Applied Technology Model Project' and a 'Provincial City Management Excellent Model Project'. It was also held up by the municipal government as a model for building technically advanced incinerator-power plants. Construction of the facility began on 18 December 2002, and after a six-month period of trial operation starting in July 2005, it officially entered production in January 2006. Its estimated daily processing volume was 1,040 tons, it had an installed capacity of 22 megawatts, and its annual electrical production was 130 million kilowatt hours.[17] The plant was built with investment from the municipal government. Because China did not have experience in running incinerator-power plants, a professional foreign company 'W' was hired to manage production. W specializes in waste management, waste water processing and other environmental services. It signed a ten-year operating contract with the municipal government. The government paid the company operating fees as agreed in the contract, and also supervised its operations, including environmental quality supervision.

Before W was hired, the plant was managed by the Waste Management Centre under the municipal government, but after reorganization, management was transferred to the GR Corporation, a state-owned enterprise group authorized by the city government. Originally its main product was lifts, but in recent years GR had extended the scope of its business to household waste management and other environmental services. In November 2006, through a competitive bid, it obtained a license to construct and operate the second phase of the L Incinerator and in July 2009 GR declared that the municipal government has granted it the exclusive licenses for the city household waste management. GR was also the company that constructed the Panyu Waste Incineration Plant, and it attracted a lot of attention and suspicion from the public.

Controlling the temperature of waste incinerator furnaces and the length of time gas remains in the furnace are important for controlling air pollution emissions. If the temperature is too low, dioxins will be formed; a

17 Information provided by the power plant.

temperature range of 300-500°C produces the highest level of dioxin. As the temperature rises, dioxins decompose and at 773°C they are almost entirely gone. To prevent the formation of dioxins, state regulations stipulate that the internal temperature of waste incineration furnaces must be higher than 850°C, and gas must remain in the furnace for at least two seconds. We observed the temperature values for the furnaces at L Incinerator on the display monitor in the control room, and they fluctuated between 950°C and 1,100°C. The staff explained that the gas pipes were designed with curves to ensure that the gas remains for two seconds and dioxins decompose.

Another criticism of waste incinerators is the way in which fly ash is treated. For example, Zhao Zhangyuan, who is widely regarded as an anti-incineration expert, is convinced that there is no way to deal effectively with fly ash from waste incineration. We visited the area of L Incinerator where fly ash was converted into solids. Because the ash produced during incineration contains a relatively high level of dioxins, it is collected and processed as hazardous waste and must be rendered leak, dust, and rain-proof. The ash at L Incinerator received proper treatment, which involved collecting it from the ash tray and adding cement and chelating agents to form a solid block. This would prevent any heavy metals, dioxins, and other substances in the ash from entering the environment and creating secondary pollution. When a certain quantity of cement blocks had accumulated, they were transported to a special location in the waste landfill for storage (as shown in Photograph 22). Our team visited and observed around ten incinerator power plants in several provinces and L Incinerator was the only one that had a process for converting fly ash into a solid form in this way.

Overall, the interior of L Incinerator was clean, there was no obvious garbage odour, and the environmental situation seemed quite good. It can be said to be one of the best incinerator-power plants that we have visited. The operation environment we observed was significantly different from the situation discussed online or described by villagers. The managers told us that even though the plant operates according to standards, because waste incineration technology has still not earned the trust of the public, many people are not familiar with it and do not support it. No matter what the enterprise says or does, the villagers are still suspicious of the technology and the enterprise. This makes communication very difficult. The managers had taken a series of measures to increase the transparency of the enterprise and gain public trust, including regularly accepting visits from government, educational and social organizations, and all kinds of other people. Special villager supervision teams had also been set up, consisting of representatives elected by the villagers themselves. These representatives had work permits and could go to the plant to investigate at any time, visit the central control room and ask employees for statistical data. However, villagers were not well versed in waste incineration equipment and technology, and so even if they went to the plant, they would only be able to see the superficial situation. The enterprise had also installed real-time monitoring equipment, and the data was presented on a monitor in front of the plant, so villagers could watch it any time. But when we asked villagers about this real-time monitoring, they said that they did not trust it. They believed that it was unlikely the enterprise would make the real statistics public for the outside world to see.

As for media reports and the suspicions of netizens, L Incinerator employees also had explanations for us. Regarding the 'rope, fabric, plastic bags, sport shoes' and other debris seen by a China Newsweek reporter, their explanation was that the waste incineration furnace had a burning reduction rate, which means that a small percentage of unburnt garbage was allowed. Unsorted waste is complicated to handle because if it is not all loose, it is not always possible to incinerate it completely the first time around. For example, some garbage might have been bound tightly with rope, or there might be rocks that are not combustible. The garbage tumbles around continuously in the furnace, but sometimes it will not fall into the right position, or sometimes the machine is faulty and would not push it close enough to the heat source to be incinerated. In this situation the incineration process would be incomplete, but the workers would retrieve the unburnt debris to be incinerated again.

Employees believed that the main reason villagers were against the plant was because of the second phase of the incinerator (L Incinerator Phase II) that was constructed next to the existing facility.

> Our operations are actually very standard. [...] before the second plant was built, there were very few complaints from villagers, not even one or two each year. And even when there were complaints, they were all about the stench, but that was not produced by us over here, it was produced when the garbage trucks came in. When the second plant started, the situation changed. They said that there was going to be additional construction here, and so we must compensate them. [...] As far as I know, the government already initiated a series of compensation measures for villagers; it is not only money, they also compensate them by giving them land elsewhere. The government said they can collectively relocate the villagers, but they do not want to move. [...] We feel that Chinese citizens are still not willing to accept waste incineration. It is really hard to build an incinerator-power plant; the voices against it are very loud, and our social responsibility is very heavy. Everyone is watching us.' (L Incinerator employee, December 29, 2010)

The same employee also mentioned the issue of supervision:

> Government supervision is very strict. Sometimes our furnace temperature does not reach the standard, and we will add more fuel, but the furnace temperature may not rise that quickly. If it does not return to 850 degrees within the regulated time, then we will be fined. But incineration is a process, and sometimes the garbage contains a lot of rocks, so no matter what we do it will not burn, and we can't do anything about it. The owner of the plant knows this. They know that garbage has particular qualities, but the EPB does not care about this. The standard is 850 degrees, and if we cannot reach it there will be a fine, no matter what excuses we have. (L Incinerator employee, 29 December 2010)

After visiting the whole plant, we happened to hear in conversation with employees that GR Corporation had also set up a special team to provide supervision and management for L Incinerator. The staff later took us to the team's office on the second floor and the deputy leader agreed to be interviewed. He explained that the team was only set up a little more than a year before and was mainly responsible for the production process and security. The enterprise commissioned a third party to conduct more comprehensive monitoring of the surrounding water and soil. The supervision team was stationed at the plant long term in order to keep on top of operations at all times and resolve any problems that might arise.

Through observation at the plant and interviews with employees, we gained a better understanding of the situation. Our overall impression was that the plant had quite comprehensive systems in place, and supervision was well organized. Although villagers said that the plant emitted more black smoke at night, we did not observe this ourselves. However, it should be noted that this was the situation we observed in December 2010. We have no way of knowing how the enterprise was operating earlier.

Government Declares 'Supervision in Place'

We decided to visit the relevant departments to learn about government regulation of L Incinerator. First, with considerable difficulty, we located the district EPB, but we were surprised to be given the cold shoulder. The staff told us that L Incinerator was under the direct supervision of the city EPB, and that they knew nothing about it. We then went to the city EPB, and sought out the environmental monitoring team, and the solid waste department, but no one was willing to talk to us. Later, a leader in the Policy and Regulation Department had a brief exchange with us and we learnt that L Incinerator was co-regulated by the owner, the industry, and the EPB.

Owner supervision referred to the supervision team organized by the state-owned GR Group; industry supervision referred to the Waste Management Centre under the Municipal Management Committee, and environmental supervision was the on-site supervision personnel assigned by the EPB. This EPB team is stationed at the plant and can check the control panel in the central control room at any time. The EPB official said that the plant's level of supervision was stricter than national standards required, and inspections were more frequent. The state required inspections once per season, but they inspected two or three times. He also explained that there were two main methods of supervision: one was the EPB's supervisory monitoring, and the other was that the enterprise itself was required to contract a qualified third party to conduct monitoring. Real-time monitoring technology had also increased the efficiency of the EPBs' supervision. They could monitor day and night, and if anything exceeded the standard, it would show in the data. However, when it came to dioxin, which the public was so worried about, current technology still could not achieve real-time online monitoring.

In addition to ordinary environmental protection supervision, the city EPB also assigned an on-site supervision group, mentioned above, to conduct direct third party supervision. The personnel had already been at the plant 24 hours a day for around half a year. They had also invited villagers to join in supervision of the plant. In terms of dioxin detection, the EPB official said that the enterprise contracted a third party to test for it at least once or twice a year, and that the government also conducted random sampling at least once a year. In terms of the management of fly ash, the third party monitoring company provided a professional testing report to show whether or not this was in compliance with national standards. When visiting villagers in Y Village, we also noticed that there were signs saying 'environmental monitoring phone number' hanging on the outside of some houses. If the operations of L Incinerator were affecting the environment, villagers could call this number to contact the relevant government departments.

Afterwards, we also went to the city CDC, which had published a report on medical examinations and household surveys of the incidence and mortality from cancer among Y villagers that it had conducted with the assistance of the Y Village Committee. In our conversations with villagers, we found that they did not trust the CDC's investigation at all. Village cadres also reported that, at the time, the CDC people just came with the cancer name list from the internet and sought out people from the village committee to learn about the situation of cancer victims and remove erroneous information. They did not conduct a full investigation. Staff at the CDC insisted that no matter what doubts people might have, their position on the matter was neutral, and their investigation was objective and fair. However, they did not wish to respond further to our enquiries and stressed that the final statistical report from the investigation had already been published in the media.

Publishing the results of its expert investigation in a timely manner was a responsible action on the part of the CDC, but one could argue that the agency had an additional responsibility to give the public a more detailed explanation of the results. Ordinary people may not have a comprehensive grasp of specialized medical information and may distrust government. Their understanding of the situation may also be mixed up with their own interests and demands. If the authorities concerned can give detailed explanations to the public, it can help to reduce misunderstandings. We have experienced ourselves how difficult it can be to communicate with the complex array of bureaucratic agencies involved, and if even researchers led by a university professor, who want to learn the truth, find it so difficult, imagine how much harder it must be for ordinary people to get information. The failure of the relevant government departments to provide timely and effective responses to internet rumours and false media reports is one of the most important reasons why the 'List of Y Village Cancer Deaths' continued to ferment on the internet.

The High Incidence of Lung Cancer was not Logical

In the winter of 2010, after several investigations in different parts of the country, we gained new insights into the economic interests, social structure, and regional cultural factors behind the phenomenon of 'cancer villages' in China. In 2011, our research for the China Association for Science and Technology gave us an overall understanding of the technology and structure of city household waste management in China and from this and our reading of the literature, we reached some basic conclusions about this issue. However, there were still important questions we could not explain. For example, why did people consider L Incinerator, which appeared to be operating normally, to be so terrible? And if there was a high incidence of cancer, were there patterns to its spatial and temporal distribution? As we pondered and discussed these questions, one suspicious detail became clearer and clearer: if the high incidence of cancer was caused by toxic emissions from the waste incineration plant, then it should have started to appear after the waste incineration power plant had been operating for some time. Because cancer is a slow onset disease, there should be a time lag between the power plant beginning operations, the appearance of pollution, and the high incidence of cancer. But the list of cancer victims on the internet and in the media showed that the high incidence of cancer started very soon after the plant was built, which does not conform with the logic of slow onset. Perhaps the cause of the high incidence of cancer was the waste landfills in the past, but if so, then accusing the waste incineration power plant was unfair. We therefore gradually identified the 'high incidence of lung cancer' as the next focus of our investigations.

In June 2011, one of the co-authors of this chapter, Li Qi, went back to gather more information by talking to the family members of 'lung cancer' victims. Cancer is generally a taboo topic in rural areas, and villagers believe it to be an 'evil disease'. Since discussing death is also a painful experience to begin with, initially neither the victims' family members nor their neighbours were willing to discuss the situation regarding cancer or deaths. Li Qi was refused entry many times, and some villagers even threatened to call the police, saying they had been 'harassed multiple times' by the researcher, and shooed her away. But her persistence finally paid off and, in the end, many residents cooperated. They suppressed their emotional discomfort, and told the researcher everything they knew. Li Qi then conducted further one-by-one verification of the lung cancer victims on the list of victims who died after Y plant was built' (see Table 8.2). The results are as follows:[18]

1. The names on the list were all people from Y Village. Apart from the age of Xu*Ying which was actually '73' but was accidentally given as '53', the information for the rest of the villagers regarding name, gender, community, and age was basically correct.

18 The verification of cancer types involves expert medical knowledge, and as social scientists, we were very cautious about this. When investigating door-to-door, we first let the victim's family members provide a medical history. However, because it is the local tradition to burn all possessions, including the medical history, and related materials after a patient dies, Li Qi was basically unable to obtain any evidence of medical histories. She could only obtain information about the diseases patients had through conversations with their families. During these conversations, she adopted certain interview techniques to verify the type of disease with the patient's family multiple times and inquired in detail about the nature of their illness and its treatment. The result of this was that the type of disease that every patient had was confirmed or eliminated by the patient's family members. Li Qi further verified the information with their neighbours, villagers who were familiar with the situation and other methods. While this was not an epidemiological investigation, if even death caused by a car accident is marked as 'nasopharyngeal cancer', then something else is obviously going on.

2 Out of 27 cases of lung cancer patients, eleven cases were verified and twelve cases were eliminated. Four cases could not be confirmed because the two of the families had moved away, and the other two families did not cooperate with the investigation. For details see Table 8.4.
3 Of the 42 victims said to have died after the plant was built, there were another 15 cases in addition to the 27 cases of lung cancer. The investigation only confirmed the status of six cases. One had correct information – Chen*Jia, who had nasopharyngeal cancer. The other five cases all had false information': Zhou*Zhen had 'bronchitis', but was listed as having 'liver cancer'; Chen*Feng had 'nasopharyngeal cancer', and was marked as having 'throat cancer'; Liu*Huang had 'oesophageal cancer', but was listed as having 'nasopharyngeal cancer'; Zhuang*Sun died in a car accident, but was listed as having 'nasopharyngeal cancer'; Zhou*Yang had 'liver cancer', but was also listed as having 'nasopharyngeal cancer'.

This investigation showed that the widely disseminated list of Y Village cancer victims contained quite a lot of misinformation, and especially in the case of the two respiratory system cancers – lung cancer and nasopharyngeal cancer –there was reason to suspect that information had been falsified. Many details in the process of our investigation revealed that the creator of the 'cancer victim list' did not conduct fieldwork. For example, in the case of the patient Xu*Ying, the actual age and the age listed were twenty years apart, and her son Mr. Zhou confirmed that he had never told anyone in the village that his mother had cancer, so it was very strange to him that an outsider might have found out that his mother had lung cancer. Additionally, Zhuang*Sun, who died in a car accident and Zhou*Lang, who died of a myocardial infarction were also listed as having died of cancer.

Discussion

The dissemination of this 'cancer name list' had a big social impact. Even though we could not find out exactly who created the list and for what purpose, events that occurred before and after this incident and other background information offer some clues as to what lay behind it.

Why did villagers oppose the incinerator power plant?

If we put ourselves in the position of the local villagers, it is reasonable that people living around L Incinerator would oppose the building of a

Table 8.4 Confirmation of 'lung cancer' deaths after the plant was built

Name	*Community*	*Gender*	*Age*	*Situation of disease*		
				Record on internet name list	*Investigation result*	
Zhou*Heng	3	Male	51	Lung cancer	Eliminated	Liver cancer
Xu*Gui	5	Male	67	Lung cancer	Eliminated	Brain tumour
Xu*Rong	5	Male	75	Lung cancer	Eliminated	Stomach cancer
Zhang*Rong	8	Male	85	Lung cancer	Confirmed	Lung cancer
Chen*Qing	8	Male	70	Lung cancer	Confirmed	Lung cancer
Chen*Ying	9	Female	56	Lung cancer	Eliminated	Rheumatic heart disease
Chen*Sheng	9	Male	63	Lung cancer	Eliminated	Liver cancer
Fan*Rong	10	Male	48	Lung cancer	Confirmed	Lung cancer
Fan*bin	10	Male	61	Lung cancer	Confirmed	Lung cancer
Xu*Ping	11	Female	49	Lung cancer	Confirmed	Lung cancer
Fan*Xin	11	Male	72	Lung cancer	Confirmed	Lung cancer
Fan*Qin	11	Male	70	Lung cancer	Confirmed	Lung cancer
Fan*Fa	12	Male	58	Lung cancer	Unconfirmed	Moved away
Fan*Biao	12	Male	47	Lung cancer	Confirmed	Lung cancer
Fan*Cai	12	Male	70	Lung cancer	Confirmed	Lung cancer
Chen*Hao	13	Female	53	Lung cancer	Confirmed	Lung cancer
Chen*Yuan	13	Male	67	Lung cancer	Eliminated	Liver cancer
Xu*Ying	13	Female	73	Lung cancer	Eliminated	Liver cancer
Ye*Wei	14	Feale	65	Lung cancer	Unconfirmed	Refused visit
Zhou*Qian	14	Male	74	Lung cancer	Eliminated	Stroke hemiplegia
Zhou*Ling	14	Male	69	Lung cancer	Eliminated	Emphysema, bronchitis
Zhou*Lang	14	Male	59	Lung cancer	Eliminated	Myocardial infarction
Zhou*Tong	14	Male	37	Lung cancer	Eliminated	Brain tumour
Zhou*Cang	14	Male	69	Lung cancer	Unconfirmed	Refused visit
Zhou*Cang	14	Male	66	Lung cancer	Confirmed	Lung cancer
Zhou*Jiu	16	Male	58	Lung cancer	Eliminated	Nasopharyn-geal cancer
Zhou*Ming	17	Male	72	Lung cancer	Unconfirmed	Moved away

Data source: statistics the author gained through door-to-door investigations

second waste incineration power plant. The villagers disliked the current L Incinerator because of the environmental damage previously caused by the landfill, and they were also worried about the increasing risk from the Phase II L waste power plant that was being built.

Construction of the landfill began in 1990, and it covered 518 *mu* of land. It went into operation in 1992. By 2000 it was full, and in 2005 it was closed down. When L reservoir was original chosen by the city government as the site for a waste landfill in 1989, the project met with strong opposition from villagers. Some of them even stormed onto the construction site to destroy machinery and stop the work. However, this opposition was eventually 'quelled'. Villagers reported that during the period when L landfill was operating, it severely polluted the surrounding environment. The areas around the landfill stank terribly, and it bred flies and mosquitoes. When it rained, polluted water from the dump ran all over the roads, fields, rivers, and fish ponds. Mr. Zhang, who had worked at the landfill, said that the choice of site was a problem, because the reservoir already had a problem with leaks, and the leak-prevention technology of the landfill was not up to standard, so polluted water from the landfill seeped into the ground water. He also confirmed that the landfill was badly managed. Because the collection and disposal of liquid that leached from the landfill was not standardized, and there were random and secret discharges. This was particularly common on rainy days when the waste water processing tank did not have enough capacity, and the plant could only discharge the water directly.

> In the past villagers all drank well water and every household had a well. There were two reservoirs on the hills around the village: R reservoir and L reservoir. The water that came down was very clear. It could irrigate nearly a thousand mu of fields and orchards in the village. Less than two years after the landfill was built, waste water from the landfill gradually got mixed into R reservoir, and it is now completely dead. No fish can survive in that water anymore. Later, the well water became undrinkable. At first, we could not see anything, but there was an odd smell, and then slowly the water turned rather dark. The village collectively installed tap water. But especially when it rained, the polluted water ran all over the road and fields. The fish and shrimps in the streams and ponds all died. The owners of the ponds went to ask for compensation. If it was dead ducks and fish, we had evidence, and we could get compensation; but there was nothing we could do about the polluted fields and water. (Villager from Y Village Group 12, June 16, 2011)

By 2000, L landfill was basically full. Villagers hoped that after the landfill had been covered and closed, the environment would gradually improve. However, the city government then built a waste incineration power plant right next to it. Because villagers had suffered the last time they opposed the landfill, when the incineration plant was being built, they just protested by staging a blockade and sit-ins. They did not dare take extreme measures to obstruct the project.

The villagers found the new Phase II plant even harder to accept. It was situated on the spare land next to L Incinerator and the closed L landfill. With a planned daily processing capacity of 2,000 tons, it occupied an area of 70,000 square metres. The plant was directly financed and operated by the Guangri Corporation. Even though villagers expressed their opposition many times, they were always stopped, and construction continued.

The public has a hostile mentality towards incineration power plants. The existence of facilities such as waste landfills, funeral parlours, nuclear power plants, crematoria etc. has a negative impact on the living environment and psychological wellbeing of local residents, provoking feelings of revulsion and a NIMBY attitude. This is not unusual. We are all creating garbage every day and people do not oppose the government building waste treatment facilities as long as they are not near their own homes. But as soon as they are located close to us, we do everything we can to oppose them. We need to find a way for the government to find a balance between benefiting the majority and protecting the minority whose interests suffer; and see how we can use public wisdom to solve this thorny problem.

How the 'High Incidence of Lung Cancer' Fuelled the Anti-Incineration Movement

Starting around 2008, more and more people gradually began to oppose the incineration of urban household waste. In August 2008, residents of Beijing's Chaoyang District opposed building an incineration power plant at the Gao'antun waste landfill. In March 2009, some residents of Nanjing jointly opposed the construction of an incinerator at Tianjingwa. In April 2009, Shanghai residents near the Hongqiao waste incineration plant opposed the expansion of the plant by 'taking a stroll'. In August 2009, the 29th International Dioxin Conference was held in Beijing, bringing the toxicity and hazards of dioxin to public attention. The controversial topic of incineration power plants began to brew, stirring up further anti-incinerator movements. In October 2009, during its test operation, the Jiangsu Province Wujiang Waste Incineration Power Plant encountered opposition from residents, and the project had to be stopped. Similarly, in March 2011, residents of Wuxi in Jiangsu Province demonstrated against Wuxi Xidong Waste Incineration Power Plant, causing the plant, which had already been constructed, to halt operations.

We cannot know who concocted the 'cancer victim list' in the case of L Incinerator and led the public to reach the 'scientific conclusion' from the analysis of this data that 'waste incineration power generation was causing a high incidence of cancer'. We did not have the capacity to investigate the full process of creating the list and the purpose behind it. But one thing seems very clear, which is that by presenting 'scientific evidence' like this on the internet and in the media, it was possible to make the public and the government conclude that toxins produced by waste incineration can be life-threatening, and this was sufficient to mobilize the anti-incineration sentiments of people who lived near planned facilities and caused them to be halted or relocated to other regions. Objectively, it has also forced those who support waste incineration to take seriously the social and health risks it may cause, and to be more comprehensive and cautious in their deliberations when planning and choosing sites for future projects.

Works Cited

Balshem, Martha. *Cancer in the Community.* Washington and London: Smithsonian Institution Press, 1993.

Chen, Jianjie. 'L, We are Sad for You', *China Newsweek* 2009 (issue 44).

Feng, Shizheng. 'The Silent Majority: Chaxu Geju and Environmental Action in China'. *Journal of Remin University of China* 1 (2007): 122-132.

Funabashi, Harutoshi. 'Minamata Disease and Environmental Governance'. *International Journal of Japanese Sociology* 15, no. 1 (November 2006): 7-25.

Guangzhou Daily. 'Y Village cancer Victim Name List Rumour Inaccurate', 26 January 2010 (page A7).

He, Yanling. '"Chinese Style" NIMBY Conflict: Analysis Based on Events'. *Open Times* 2009 (issue 12): 102-114.

Jing, Jun. 'Consciousness and Awareness: Environmental Resistance of a Xibei Village'. *Journal of China Agricultural University (Social Science Edition)* 2009 (issue 4): 5-14.

Li, Qi, and Chen Ajiang. 'On "Fake" Data of High Incidence of Cancer, Based on Research of Sociology'. *Journal of Nanjing University of Technology (Social Science Edition)* 2013 (issue): 117-128.

Qing, Yan. 'L High Incidence of Cancer and Its Correlation and Causality with Waste Incineration Project'. *Wealth Investment Weekly* 2009 (issue 81). http://www.360doc.com/content/11/0509/19/6017453_115534291.shtml. Accessed September 10, 2010.

9 Villagers' Perceptions of and Responses to the Relationship between Cancer and Pollution[1]

Chen Ajiang and Cheng Pengli

Abstract

It is difficult for medical science to establish a clear relationship between cancer and pollution at the village level due to the small population size and the multi-causality of many cancers. But 'cancer villages' nonetheless exist as a social fact and they continue to affect the lives of villagers. In the face of health risks, and uncertainty about them, villagers try to avoid pollution by getting polluting industries to shut down or relocate, moving away themselves, and changing their water sources and diet. Economic circumstances affect the degree to which they can deploy these various strategies. Their responses are also shaped by the inherited social structure, including the power of the village clan, local 'gentry', and the work unit.

Keywords: causes of cancer, water pollution, social structure, local knowledge, village traditions

Introduction

We first encountered the phenomenon of 'cancer villages' while conducting research for the project 'On Harmony Between People and Water'. Further investigation showed that there were big discrepancies between our expectations and media reports, and the actual health problems caused by pollution. We observed that as pollution got worse, the incidence of disease would

1 This article was originally published in *Xuehai*, 2011 (issue 3).

Chen Ajiang, Cheng Pengli, and Luo Yajuan, *Chinese "Cancer Villages": Rural Development, Environmental Change and Public Health*. Taylor & Francis Group 2020
DOI: 10.5117/9789089647221_CH09

increase; but at a certain point, even as pollution continued to become more severe, the incidence of disease might start to stabilize or even decrease. We can assume that this is because when certain abnormalities begin to appear in the incidence of sickness or death, villagers take defensive measures to avoid harm, and so even if pollution continues to worsen, after reaching a peak, the rate of disease will not continue to rise and may sometimes even fall.[2] However, due to the lack of detailed empirical material, we did not pursue further research on this issue at that time.

In December of 2010, we conducted a study on environment and health in Shangba Village and Liangqiao Village in Shaoguan, Guangdong; Jiannan Village in Shangrao, Jiangxi; and Xiqiao Village in Jiaxing, Zhejiang.[3] These four villages all suffered from relatively severe industrial pollution, the incidence and mortality rates of cancer were abnormal, and the villages were referred to by local people or the media as 'cancer villages'. The use of the term 'cancer village' here is therefore borrowed from the media or residents and is not a scientific term. We had previously conducted multiple rounds of fieldwork in Xiqiao Village in Jiaxing, Zhejiang, Dongjing Village in Yancheng, Jiangsu, and Huangmengying Village in Zhoukou, Henan for the project 'On Harmony between People and Water' and so there is a connection between that project and the research we discuss here.

The surface and ground water in the villages of Shangba and Liangqiao in Guangdong was contaminated with heavy metals from many years of mining activity upstream and there was an abnormally high incidence of disease. This resulted in a lot of media reports, and also in the involvement of scientists and NGOs. Jiannan Village in Jiangxi was originally a state-owned farm, and at the time of our research still had certain characteristics of a 'work unit', reflecting the special economic and social system of such farms. Although Jiannan by this time resembled a city district; for ease of comprehension, we continue to refer to it as a village. The fourth village discussed here, Xiqiao in Zhejiang, is quite a typical example of the sharp contradiction between environmental pollution and public health that occurred during the process of industrialization in rural areas of northern Zhejiang, southern Jiangsu, and other regions with similarly rapid industrial development. For background descriptions of the four 'cancer villages' see Table 9.1.

For pragmatic reasons, our team used mixed research methods. We first gathered information on pollution and disease in the research sites by reviewing the literature, including natural science articles. Research and

2 Chen, 'On Harmony between People and Water', pp. 100-101.

3 In accordance with academic practice, the names of the villages have been changed.

Table 9.1 The basic situation of the four 'cancer villages'

Background	*Shangba and Liangqiao*	*Jiannan*	*Xiqiao (Dingbang Natural Village)*
Overview	Shangba and Liangqiao are two neighbouring administrative villages in the same town. In 2005, Shangba had a population of 3,329 people and Liangqiao a population of 382. The two villages are situated in the North Guangdong mountain area. Its level of socio-economic development was in the lower middle range for Guangdong Province. In 2010 the average rural income of the county in which the village was located was 6,138 yuan.	Jiannan has a population of around 1,000 residents who are all workers of the state-owned farm. Because the farm had not made much profit in recent years, the income and living conditions of the villagers were quite bad. In 2010, the average rural income of the county in which the village was located was 4,482 yuan.	Dingbang is a natural village in Xiqiao administrative village, which includes villager Groups 4 and 5. In 2005 its total population was 192 people. Local rural industry was relatively developed and the average income of villagers was quite high. In 2009 the average income of the district's rural population was 12,531 yuan.
Pollution sources	Waste water from state-owned and private mining enterprises, containing lead, zinc, copper, cadmium, and other heavy metals	Historically, there was a Monosodium Glutamate (MSG) factory, and a caustic soda plant which had both closed. Villagers suspected that the glass factory and the phosphate fertilizer plant produced pollution, but there was no evidence to support their suspicions, The well water in some villagers' homes is heavily alkaline, and lead, zinc, cadmium and other substances exceeded standards.	The waste water discharged by the metal furniture factory contained sulfuric acid and unknown pollutants; the white waste water discharged by water jet textile mills also contained certain chemical substances.
Disease	Between 1986 and 2005, Shangba and Liangqiao had an annual average of 214 and 23 cancer deaths, respectively. The annual cancer mortality per 100,000 people for the two villages was 315 and 301, respectively.	Between 2000 and 2009, a total of 32 people died from cancer in Jiannan. The annual cancer mortality rate was 320 per 100,000.	Between 2000 and 2007, nine people died of cancer in the two villager groups in Dingbang, equivalent to a cancer mortality rate of 586 per 100,000 people.
Villagers' perceptions and the relationship between cancer and pollution	Research has proved that toxic substances discharged from upstream mines contain heavy metals. The villagers' drinking water and crops also contained high levels of heavy metals. The pollution-disease relationship is clear.	Villagers believed the glass factory discharges toxic substances leading to a high incidence of cancer. Testing of the villagers' drinking water did not show a clear link to the pollution from the glass factory. The pollution-disease relationship is unclear.	Villagers believed the metal furniture factory discharges toxic substances leading to a high incidence of cancer. But there has been no testing or the results have not been made public. The pollution-disease relationship is also unclear.

test reports conducted by government agencies, civil society organizations and individuals, and 'cancer death lists' compiled by villagers themselves also helped us to understand the situation. Field observations were essential in enabling us to make an overall assessment of the topography of the village and the economic and social situation of the community, as well as giving us a sense of the pollution situation on the ground. Semi-structured in-depth interviews were our most important source of information. Based on experience from past research, we adopted a bi-directional approach, in which we both 'went down' to the villages, to chat and discuss with residents, and also 'went up' to the government Environmental Protection Bureau (EPB), Centre for Disease Control (CDC), and other agencies, to learn about their understanding and management of pollution and disease.

From the perspective of medical science, the relationship between pollution and disease is very complex; there is a strong element of uncertainty and many possible scenarios. In many of the 'cancer villagers' on which the media has reported, the relationship between pollution and disease is in fact not clear at all. In some cases, more time is needed to accurately conclude whether there is a cause and effect relationship, and in most cases, it is actually unlikely that any final conclusion will be reached. Even though many 'cancer villages' have been neither confirmed nor falsified by scientists, the existence of 'cancer villages' is nonetheless a social fact that has already shaped the public's perception of environmental health issues and exercised an irreversible influence on responses and institutional policies towards them. From the perspective of sociology (or broadly speaking from the perspective of the social sciences), it is therefore meaningful to analyse and describe the processes of how and when pollution occurs and how and when abnormalities appear in the incidence of disease and mortality; how villagers understand and interpret these phenomena; how they understand the relationship between these uncertain variables; and how they respond in the context of unclear scientific evidence.

Interpreting the 'cancer-pollution' relationship

After careful consideration, we found that the degree to which the 'cancer-pollution' relationship can be determined does not usually fall at either of the extreme ends of the spectrum from 'clearly known' to 'completely unknown', but rather somewhere in between. Based on this assumption, and for ease of expression, we divide this spectrum of knowledge into gradually decreasing levels of certainty: (1) certain knowledge, where the relationship between cancer and pollution has been proven; (2) speculation, based on considerable evidence of a relationship; (3) speculation, based on relatively little evidence of a relationship; and (4) complete uncertainty, or speculation without any proof.

Historically, situations in which the relationship between pollution and disease has been confirmed are rare. Minamata disease in Japan is one of the few cases in which the causality was clear. But even then, where there was scientific proof that methyl mercury caused Minamata disease, recognition of this fact and compensation for sufferers of the disease was still the subject of a prolonged dispute.[4] Our research on 'cancer villages' in China shows that it is difficult even for authoritative government agencies to determine the 'cancer-pollution' relationship and so it would obviously be even more of a challenge for villagers. This explains why, even though the issue of 'cancer villages' has been widely reported in the media, and circulated on the Internet, and it has generated a strong reaction from the public, government agencies have still responded with a vague and ambiguous attitude, making the issue ambivalent and uncertain.

The uncertain scientific relationship between cancer and pollution means that it is very challenging to conduct social science research on the topic and we have sometimes wondered whether social scientists should perhaps put it aside until a more promising way of working on the issue presents itself. But in certain contexts, inaction is itself also a certain kind of action. Even though the scientific relationship to pollution may not be clear in the case of any particular 'cancer village', people affected by disease will have suffered and people exposed to pollution will also have responded in some way. The legal system is therefore faced with a situation that is both pressing and at the same time very hard to deal with. For example, in some cases, villagers have sued chemical plants for causing their health problems. In this situation, if the court decides in the villagers' favour, there is often a problem of inadequate proof in terms of the evidence provided and what is required by law. Yet, if the court decides against the villagers because the pollution-disease connection is unclear and there is insufficient evidence, there will be other, social, consequences. Inaction on the part of the court is, *de facto,* a form of action that condones the polluting behaviour of the chemical plant, and this situation is somewhat ironic if the law is supposed to function as a mechanism of justice.[5] Put another way, scientists can say

4 Funabashi, 'Minamata Disease and Environmental Governance', pp. 7-25.

5 Japanese legal experts have already made advances in this respect, for example, in addressing problems of causality in epidemiological theory, probable cause theory, and indirect disproof

'objectively' that they 'do not know', and judges can cautiously point to the fact that there is 'insufficient evidence', but these statements all embody particular meanings and values, and they all contribute to a trend that will have a series of social consequences.

It was in this context that we became involved in 'cancer-pollution' research. The difficulty of the topic for social scientists means that the choice of approach is critical. On the one hand, we firmly believed that we must pay attention to natural science research. Social science researchers should make full use of existing natural science findings and clarify scientific problems and relationships to the best of their ability. We then sought to draw on the strengths of sociology, making comprehensive use of different research methods and conducting in-depth fieldwork. We collected a wide range of information from different sources and sought to learn from people. If key variables and the relationship between them were not clear, we tried to determine the probable range of variance to minimize errors of perception. At the same time, we stressed the need to show understanding towards people involved in the situation, and as a result, we were able to appreciate how residents and government departments viewed these matters, to what extent their attitudes were similar, and when and why they differed.

Below we analyse the nature of the relationship between cancer and pollution in the context of the four cases.

Shangba and Liangqiao Villages

Of the four cases, the pollution situation in Shangba and Liangqiao is the clearest, and this is also the case in which there is the most certainty about the 'pollution-cancer' relationship. On the spectrum we laid out above of the degree of certainty regarding whether 'pollution causes cancer', this case falls into the category of 'speculation, based on considerable evidence'.

We were not able to obtain data on the incidence of cancer or mortality in Shangba. The Village Committee had a list of villagers who had died from cancer over the years, but it refused to give us this list (for reasons explained later). However, there had been numerous media reports about cancer deaths in Shangba and, combining information from the CCTV show 'Half

theory. These approaches demonstrate the new challenges to the Japanese legal community stemming from the complexity of causal relationships and the infringement of environmental rights and the innovations they have made in response. See Ding, 'On Foreign Environmental Rights Infringement Causality Relationship Theory – Improving China's Related Theories', pp. 96-100.

an Hour on the Economy' on 8 January 2001 and 3 June 2005 with a report from 'Nanfang Metropolis Daily' on 18 November 2005, we estimated that between 1986 and 2005, 214 people died of cancer in Shangba. The village had a total population of 3,400, which means that the annual number of cancer deaths was 315/100,000 people, or more than twice the average rural cancer mortality rate. Liangqiao provided us with a list of cancer deaths which showed that 23 people died from cancer between 1986 and 2005. In 2005 the village had a population of 382 people, so the number of cancer deaths per year was 301/100,000 people, or more than twice the average rural cancer mortality rate.

The pollution in Shangba and Liangqiao originated from mining activity upstream, including the provincial state-owned B Mine company and some illegal private mining. B Mine is a multi-metal mine, primarily producing iron ore, but also copper, tungsten, bismuth, molybdenum, gold, silver, and other metals. Although the company built tailings ponds downstream from the mine, the stored waste water constantly flooded over the edge of the dam and into the rivers below. Waste water and sludge passed through two small rivers that finally merge at Liangqiao. This affected the drinking and irrigation water of downstream residents. Data from the Second National Agricultural Soil Survey shows that in areas affected by pollution in Shangba, the level of lead in agricultural soil was 44 times the national standard and the level of cadmium 12 times. The pH was around 4. Natural scientists from institutes of higher education in Guangdong Province conducted repeated tests of the water and soil in Shangba and a few other heavily polluted villages and found similar pollution issues. In addition, Zou Xiaojin and others tested the level of heavy metals in Shangba's soil and villagers' well water, rice, and vegetables, and found that multiple types of heavy metals were severely in excess of the standard. For detailed data see Table 9.2.

Research has shown that lead, chromium, cadmium and arsenic can cause a variety of cancers in humans. The polluted water bodies and crops of Shangba and Liangqiao contained lead and cadmium at levels severely exceeding national standards. Heavy metals enter the human body through two main pathways: drinking water and food that has been grown on polluted land. The ingestion of heavy metals through rice, vegetables and other foods was the main health risk to local residents, and rice was the main source of exposure.[6] By analysing the incidence of cancers of the digestive tract in Shangba and the situation of water pollution near the

6 Zou, Qiu, Zhou, et al., 'Research of Health Hazards of Dabao Mountain Mining Area Heavy Metal Pollution'.

Table 9.2 Levels of heavy metals in Shangba's soil, well water, rice, and vegetables

Subject tested	*Values*	*Pb (Lead)*	*Zn (Zinc)*	*Cd (Cadmium)*	*Cu (Copper)*
Soil	Average value of samples tested (mg/kg)	843.7	2112.0	2.66	586.3
	National soil quality standard 2 (GB15618-1995) (mg/kg)	250	200	0.3	50
	Comparison **	3.4 times	10.6 times	8.9 times	11.7 times
Well water	Average value of samples tested* (mg/kg)	0.33 μg/L	0.60 mg/L	4.61 μg/L	0.04 mg/L
	National drinking water quality standard (GB5749-85)	50 μg/L	1 mg/L	10 μg/L	1 mg/L
	Comparison **	Did not exceed	Did not exceed	Did not exceed	Did not exceed
Rice	Average value of samples tested* (mg/kg)	2.81	60.83	0.58	6.54
	National food safety standards (GB2715-2005) (mg/kg)	0.20	20.00	0.20	10.00
	Comparison **	14.0 times	3.0 times	2.9 times	Did not exceed
Vegetables	Average value of samples tested* (mg/kg)	0.80	22.45	0.28	1.07
	Vegetable standard limits (mg/kg)	0.2	20	0.05	10
	Comparison **	4.0 times	1.1 times	5.6 times	Did not exceed

Notes:

* Data for the 'average of samples tested' is taken from Zou Xiaojin et al., 'Research of Health Hazards of Dabao Mountain Mining Area Heavy Metal Pollution', *Environmental Science Daily*, 2008 (issue 7). The research shows that the average level of heavy metals contained in the well water sample did not exceed drinking water quality standards, but in the 31 wells the researcher tested, concentrations of Zn in 12 of the wells and Cd in 14 of the wells were above standards.

** 'Comparison' is the number of times by which the measured values exceed or fall below the national standard.

village, Liu Yisheng and colleagues concluded that water pollution was closely associated with the high incidence of these cancers.[7] In this case, then, the facts regarding pollution and disease are relatively clear, and so is the logical connection between 'pollution and cancer'.

Our fieldwork found that villagers were aware of pollution mostly on an intuitive, emotional level. They knew that the water was contaminated, but they did not know what kinds of toxic substances were in it and they were unclear about the health effects they might have. Villagers reported that from the 1970s, fish and shrimp in rivers polluted by the mines started dying in large numbers and eventually died out completely. Villagers found that if their skin touched the river water it would become itchy and they developed red rashes. Other people spoke of a pungent and disgusting sulphur smell in the vicinity of the river, but they could not provide more detailed descriptions.

As Shangba's reputation as a 'cancer village' grew, a constant stream of media reporters, scientists and civil society organizations went to the village to conduct interviews and research. This outside attention had two effects. The first was to alert villagers to the seriousness of Shangba's status as a 'cancer village', because if the situation were not serious then it would not attract so many important figures and organizations to come and study it or provide assistance. The second effect was that outsiders' news reports and research findings helped the villagers understand the environmental problems and health hazards they were living with. Through this process, villagers' awareness of the pollution-disease relationship deepened and the measures they later took to respond to these risks, such as changing their water sources, were closely related to their perception of pollution and disease.

Jiannan Village

In Jiannan, the facts about pollution were relatively clear, and so was the situation regarding disease. However, when it comes to the relationship between cancer and pollution, even though village residents firmly believed that the high incidence of cancer was caused by pollution from the glass factory, there was no evidence of this from research by the county CDC, or from tests of water samples sent for analysis by the villagers themselves. Judging from the available information, residents' claims that pollution was

7 Liu, Gao, Wang, et al., 'Etiological Research on Malignant Tumours of the Digestive System in a High Incidence Village in Guangdong', pp. 1139-1141.

causing cancer fell roughly into the category of 'speculation, with relatively little evidence' on our spectrum.

Jiannan residents provided us with a list of cancer patients and deaths in recent years.[8] It gave the name, age at death, and type of cancer for each victim. Jiannan had around 300 households and nearly 1,000 people, and from 2000 to 2009, a total of 32 people had died of cancer, so the cancer mortality rate was approximately 320/100,000 per year, or around twice the national average. In terms of age, the youngest cancer victim was 18, five were between 30 and 40 years old, and 27 were less than 70 years old. Two points deserve special note. First, there were five households where more than two people died of cancer, and due to multiple consecutive deaths in the family, some households were no longer willing to live in the community and had sought refuge with relatives or friends elsewhere. Second, the houses of around 20 of these 32 people were clustered along the two sides of the main road in Jiannan, within a radius of not more than 200 metres. The residents contracted many types of cancer, of which liver, stomach, and blood cancer accounted for a large percentage.

Villagers attributed the high incidence of cancer to industrial pollution in the community. Initially, pollution mostly came from three collective enterprises established by the state-owned farm: an MSG factory, a soda ash plant, and a glass factory. The MSG factory was built around 1970, and for a long period of time it had good revenue, but its waste water was discharged directly without being treated. The soda ash plant was built in early 1980s and shut down two years later. While the plant was in operation there was an accident and some caustic soda leached into the groundwater; the current high pH value detected in well water could be related to this. The glass factory was built in 1987 and went bankrupt in 1995. In 1997 it was contracted out to an entrepreneur from another province. The production process was relatively simple and mostly involved forging scrap and old glass into new glass bottles for safflower oil. The residents said that the existing glass factory caused a wide range of pollution, and primarily suspected that it discharged waste water that polluted their wells. Finally, at the time of

8 Even though we do not agree with the residents' speculations regarding the pollution-cancer relationship, we do believe that this list of cancer deaths is basically reliable. We checked all the cancer victims on the list, and roughly determined the victims' year of death and where they resided while alive. We then cross-checked the information. by asking other residents near Jiannan their thoughts on the high incidence of cancer in the area. Even officials from the county EPB (as lay people not experts) believed that the local incidence of cancer was quite high. Additionally, in terms of the psychology of Chinese people, it is very unlikely that they would intentionally fabricate a list of cancer victims.

our research there was also a phosphate fertilizer plant in operation close to the Jiannan residential area, and the people who lived there believed that it also generated pollution that was harmful to their health.

From the County EPB we learnt that the phosphate fertilizer plant only mixed rock phosphate bought in from outside with other fertilizers; there were no complex chemical reactions, and it did not generate new pollutants, so the residents' suspicion that pollution from this source was causing cancer could be eliminated. The situation of the glass factory was slightly more complex. Residents claimed that minerals containing arsenic were added in production process. But township leaders responsible for environmental protection said that while the former glass factory did use minerals containing a small amount of arsenic to make the glass transparent, the amount was very small, and it was no longer used anymore. Furthermore, the water quality test reports villagers provided (and which they regard as evidence – see the data in Table 9.3), showed no problem with excessive levels of arsenic.

Water samples from the Liu residence exceeded the national standard for cadmium. Research has demonstrated that there is a connection between cadmium and cancer, and so high cadmium content could be one of the reasons behind the high incidence of cancer in the region. But oddly, the water sample from the Li residence in the same community did not exceed the national cadmium standard. Furthermore, no connection between cadmium and the production techniques of the glass factory was found. The manganese content in the water sample from the Li residence exceeded the standard. Manganese is an essential element for the human body, and a deficiency of it can cause certain diseases, but current research does not demonstrate a clear link between manganese and cancer. The research team learned at the County EPB that it is common for the manganese content of ground water in Jiangxi Province to exceed the national standard. Soil in the province has high levels of iron, and manganese often coexists with iron, so the manganese content in red soil is also often high. According to this logic, the high content of manganese in the area is probably related to the local geology, and there was no evidence linking it to pollution from the glass factory.

To further explore the residents' claim that the high incidence of cancer among Jiannan residents was caused by pollution from the glass factory, we made a special visit to the County CDC. Judging from its previous research, no connection could be found between pollution from the factory and the incidence of cancer among the residents. Nonetheless, the villagers remained convinced that cancer in Jiannan was related to pollution. The data and analysis from the Jiangxi Province Health and Epidemic Prevention

Table 9.3 Test results for three water samples from Jiannan

Index	*Water Sample 1 (Liu residence)*	*Water Sample 2 (Liu residence)*	*Water Sample 3 (Li residence)*	*National Standard Limit of year 2006*
Chroma	*20*	<5	10	15
Turbidity	*5*	*6.5*	*27*	1
Smell and taste		Has an astringent taste	Has an astringent taste	No strange smells or tastes
pH value	*9*	*13*	6.77	No less than 6.5 and no more than 8.5
Total hardness	150 mg/l	*620.6mg/l*	*560.5 mg/l*	450mg/l
NH_3-N	–	0.8mg/l	*1.6mg/l*	0.5mg/l
Iron	<0.3 mg/l	0.03 mg/l	*0.31 mg/l*	0.3mg/l
Manganese	*0.105mg/l*	0.06mg/l	*1.69mg/l*	0.1mg/l
Copper	<1.0mg/l	0.01mg/l	0.005mg/l	1.0mg/l
Zinc	<1.0mg/l	0.005mg/l	*1.08mg/l*	1.0mg/l
Lead	<0.05mg/l	*0.035mg/l*	*0.02mg/l*	0.01mg/l
Chromium	<0.05mg/l	<0.04mg/l	<0.004mg/l	0.05mg/l
Cadmium	–	*0.05mg/l*	*0.006mg/l*	0.005mg/l
Arsenic	<0.05mg/l	<0.05mg/l	<0.05mg/l	0.01mg/l
Cyanide	<0.05mg/l	<0.002mg/l	0.002mg/l	0.05mg/l
Chloride	–	*290mg/l*	*460mg/l*	250mg/l

Notes:

1. Jiannan residents took three water samples from three wells of community residents between 1999 and 2000 and sent them for testing. The testing facility for water sample 1 was the Provincial Health and Epidemic Prevention Station, but the residents thought that its test results were unreliable. They therefore sent water samples 2 and 3 to be tested at the Provincial Health and Epidemic Prevention Station. The national standard was added by the authors for comparison.
2. Italicized text in the table represents values above the national standards.

Station mentioned above, which they regarded as evidence of this, drew the following conclusions:

> According to analysis of the water samples, the pH level, turbidity, ammonia nitrogen (NH_3-N), nitrites, total hardness, cadmium, and solubility of the water sample (from the Liu residence) exceed national health standards; the other indicators are in line with the 'Drinking Water Health Standards'. The turbidity, ammonia nitrogen and nitrogen compounds, total hardness, manganese, and zinc in the water sample (from the Li residence) exceed standards; the other indicators are in line with the 'Drinking Water Health Standards'. (GB5749-85)

This result shows which indicators met the standards and which did not, but it did not link the findings to disease. Jiannan residents have repeatedly sent representatives to petition the higher authorities, arguing that these tests prove that pollution is causing disease. Mr. Zhou was one of these representatives, and an opinion leader in the community. His wife and son passed away from cancer a few years before our research began and he was convinced that the high incidence of cancer in Jiannan was directly linked to pollution from the glass factory. Zhou was basically illiterate; he knew that the water samples did not meet the standard, but he did not understand the full meaning of the test results. When we tried explaining the significance of the indicators and their relationship to pollution from the glass factory to him, Zhou displayed strong displeasure and demanded, 'Were you people sent by the EPB? If you were sent by the Bureau, then I will not speak to you any longer'. From his behaviour we realized that in addition to the complexity of the 'cancer-pollution' problem itself, and the lack of scientific knowledge among people affected by it, the emotions and interests of those taking action also affected residents' perceptions of the underlying pollution-disease relationship.

Xiqiao Village

Groups 4 and 5 of Xiqiao administrative village, which make up the natural village of Dingbang, are the groups in Xiqiao that suffered the most severe industrial pollution. Dingbang villagers believed that the cancer incidence rate in Xiqiao and neighbouring villages was especially high and knew that they were referred to by locals as a 'cancer village'. Among the cases we studied the cancer situation in Xiqiao was characterized by being small scale and of short duration. Also, even though the cancer mortality rate was the highest among the cases we studied, we did not get the impression that an intensely confrontational mood existed among residents, enterprises, and government that was discernible to outsiders. Because almost no media, research institutions or other external actors had intervened in Xiqiao, residents' actual knowledge about issues of pollution and disease was also the most limited. On our spectrum of uncertainty, this case falls roughly into the category of 'speculation with relatively little evidence'.

The northern Zhejiang region where Xiqiao Village is situated, and the neighbouring region of southern Jiangsu, are famous nationwide for their highly developed rural industry. Industrialization of the countryside drove the rapid take-off of the local economy, but it also brought severe and widespread water pollution.[9] Pollution in Xiqiao came from two types of

9 Chen, *Secondary Anxiety: A Social Interpretation of Water Pollution in Lake Tai.*

enterprise: a metal furniture factory, and water jet textile mills.[10] The previous incarnation of the metal furniture factory was a scrap metal processing factory. In 1997, it built a relatively large facility across the river from Dingbang, which at the time of our research had developed into a medium to large scale enterprise owning multiple factories. Waste water discharged from factory's operations had polluted nearby rivers. In the early days, the factory used to purchase scrap metal and process it to make furniture. It used sulphuric and other acids to clean the surface of the scrap metal and discharged the waste water into the rivers along with waste water from zinc plating. The factory also emitted waste gases and smoke. Furthermore, in the early days, the factory would frequently secretly burn solid waste from the production process, creating further air pollution. According to the villagers, from the time the metal furniture factory was built in 1997 until around 2001, waste water was discharged untreated into the river. Later, after 2005, due to the villagers' continuous opposition, the situation somewhat improved.

Dingbang is a typical Jiangnan (South of the Yangtse) village, in which the homes are arrayed along the banks of the river. The metal factory which villagers suspected of creating health risks was nestled close among them, as shown in Figure 9.1.

Between 2000 and 2007, nine people in the two Dingbang villager groups died of cancer. The total population was 192, and so the annual death rate was 586/100,000 people, or approximately four times the average rural cancer mortality rate. There were also two cancer survivors. The high incidence of cancer in Xiqiao not only occurred within a small geographic scale, but was also concentrated in Dingbang, which was right next to the metal furniture factory. The outbreak also occurred over a short period of time. The year with the highest concentration of deaths was 2003, when two relatively young people passed away. In terms of the timing, the pattern of deaths from cancer fits with the onset of pollution from the metal furniture factory and subsequent measures to reduce it. The high incidence of cancer occurred in the three- to five-year period after severe water pollution began. Later, when the villagers realized the seriousness of the problem, they took preventive measures by changing their sources of water and food, and the factory also improved its waste treatment. The incidence of cancer decreased.

10 Xiqiao Village textile mills were small scale and numerous. At the time of our research there were over 80 enterprises. When water jet looms are in operation, the size on warp yarns dissolves or falls off, ultimately entering the water. At this time, warp yarn size on the market was mainly acrylic. Although it is quite biodegradable, and not toxic to the human body, the size caused severe pollution to water bodies, and aquatic creatures living in them basically died out.

Figure 9.1 Map of the location of the Xiqiao metal furniture factory

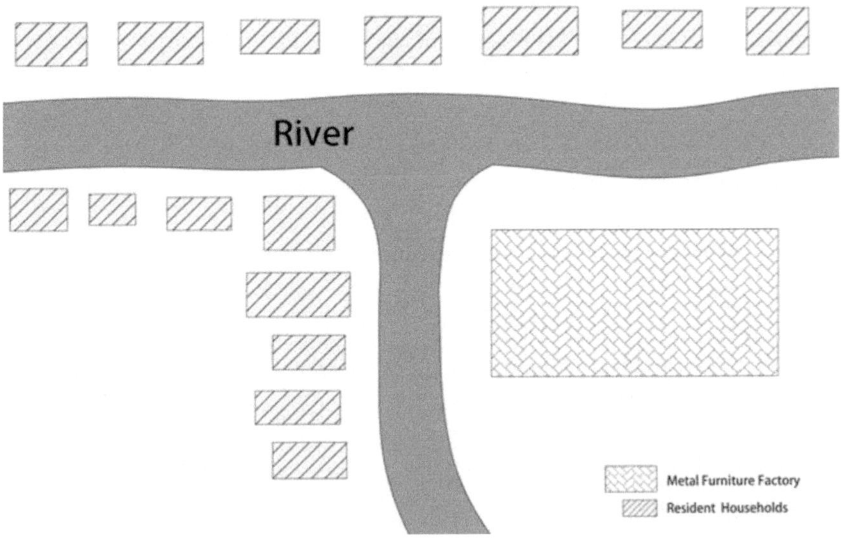

Our enquiries with the relevant government agencies found that they knew a certain amount about the severe state of pollution in the area. For example, the Jiaxing Xiuzhou District EPB had been making efforts to solve the pollution problem by building a network of waste water pipes for the textile enterprises in the village. And starting in 1992, the Jiaxing Municipal CDC had set up a monitoring system for severe diseases, including a system for reporting cancer. Ten years previously it had also established a database for major diseases. But these agencies had limited awareness of detailed, local cases such as Dingbang. When we inquired at the Municipal CDC about cancer in the village, they explained that there was a slightly higher incidence of cancer in the municipal area, but it did not exceed the upper limit and no regional differences were found. Considering the sensitivity of questions about cancer, we can also take this as a tactful acknowledgement that the incidence was high, but evidently polluted areas such as Xiqiao were still not receiving enough attention in terms of government disease monitoring.

It therefore makes sense that villagers knew relatively little about the actual situation regarding pollution and cancer. It was very difficult for them to know exactly what type of pollution factories were causing. They could not legally enter them and the enterprises deliberately concealed pollution by discharging waste water secretly, making it even harder for ordinary residents to know the truth. They sought help from the local government and the media, but both fobbed them off with euphemistic excuses. The complex production processes and professional knowledge involved made it hard for villagers to learn the truth, and with the pollution situation unclear, the complex relationship between cancer and pollution was even harder for them to grasp.

Responses to Health Risks

Although different disciplines have different definitions and understandings of the concept of risk as the result of their different research perspectives, it is generally agreed that risk has two fundamental characteristics: uncertainty and the potential to pose danger or loss. In the face of this uncertainty regarding the relationship between pollution and cancer, the possibility of enormous harm, and the absence of scientific information, villagers urgently need to find ways to reduce the health risks they face in their everyday lives.

The development of science and technology has greatly reduced the mental territory that used to be the domain of religion or common sense. When 'educated' people encounter the problem of the relationship between cancer and pollution, the first thing they think of is to turn to science. But scientists have not solved the problem for which residents of 'cancer villagers' seek an answer, and if we cannot realistically expect science to do this, then the problem must be addressed in daily life. Facing health problems that might be caused by pollution, and in the absence of the scientific clarity they hoped for, villagers could only turn to their own experience, and take preventive measures on the basis of their estimation of the level or probability of risk, measures which would, in turn, only be as effective as their assessments of the level or probability of those environmental health risks were accurate. As the preceding statement is too general, and the actual situation is very complex, we analyse some specific situations below.

Because our topic of our research was 'cancer villages', the issue of disease in the cases we studied was to some extent pre-determined (even though in fact it sometimes turned out to be uncertain). The relationship to pollution was less clear. When it appears that pollutants emitted by an enterprise are affecting the health of residents, they can be categorized as 'pollutants known to cause health risks' or 'pollutants not precisely determined to cause health risks'. In these two situations, people who are affected can adopt a number of different strategies.

In the villages we studied, there was not a single case in which villagers knew about the high incidence of cancer, and about the precise polluting substances, and in which they also understood the relationship between

Table 9.4 Ideal types of responses to health risks in 'cancer villages'

Cancer	*Pollution*	*Pollution-cancer relationship*	*Measures to counter pollution*
High incidence of cancer	Pollutants are accurately identified	Pollution-cancer relationship confirmed	Precision targeting of the source of pollution. In reality, this is rare
		Pollution-cancer relationship is uncertain	Preventing risk of disease
	It is known that enterprises are polluting but specific pollutants are not known	It can be inferred that the high incidence of cancer is related to pollution	Preventing risk of disease (scale of the risk increases, and accuracy of counter measures decreases)
		Pollution-cancer relationship is uncertain	Preventing risk of disease (scale of risk further increases, and accuracy of counter measures further decreases)

pollution and cancer. The reason for this has already been explained. In a few cases, villagers could not confirm any of these variables precisely and could only take measures against the possibility that pollution was causing cancer. These measures were mainly preventive. In terms of the categories in Table 9.4 above, the likelihood of villagers taking preventive measures increased with the uncertainty of the nature of the pollutants and the pollution-disease relationship, but at the same time, the accuracy of their response measures continuously declined.

Our case studies show that villagers took various types of action. The most effective of these, we call the 'quarantine tactic'. This entailed either getting the polluting enterprise shut down or making it relocate, or alternatively, the affected people themselves moving away from the pollution source to another area. In reality, however, a more feasible approach was usually some kind of partial intervention, which involved taking measures on the basis of an estimated range of risk and seeking ways to coexist and compromise with the polluting enterprise. In fact, villagers often countered health risks primarily by changing their sources of water and food.

Eliminating the Pollution Source

Eliminating the pollution source is the simplest way to counter health risks and the one that comes most easily to mind. It entails stopping the polluting enterprise's production by any means necessary, or forcing the factory to relocate elsewhere. The advantage of this solution is security, in that it can completely eliminate the source of the risk of disease. But it is not very accurate, it comes at a high price, and it is not easy to make this method work in real life. Especially in the early stages of market-based economic development in China, when the 'hard logic' (*ying daoli*) of economic growth was being strongly promoted by the government and widely recognized by society, the economic power of enterprises was much greater than the social power of residents. In reality, it was not easy to get an enterprise to relocate.

Once pollution had occurred, and especially when it was affecting the health of villagers, the most direct response was to demand that the enterprise stop discharging polluting chemicals. The 'three-way' approach (seeking solutions from the enterprise, the government, and the media) adopted by Dingbang villagers in an attempt to solve the problem of pollution is a classic example. In 2005, within the span of one week, two people in Dingbang were diagnosed with oesophageal cancer, both of them terminal. The villagers organized and went to the metal furniture factory to report the problem, but when they arrived, they could not find the factory director. Other leaders of the factory received the villagers, but they made all sorts of excuses to evade responsibility, and the villagers did not receive a satisfactory response. When direct interaction with the metal furniture factory yielded no results, villagers began to report the issue to the village committee and the local government. They went first to the Village Party Secretary, but his response was: 'there is nothing we can do, either.' They then sought help from the retired former Township Party Secretary, who was also a Dingbang local. The old secretary wrote a document entitled 'Reporting a Situation, Making a Suggestion', and entrusted someone to take it to the director of the metal furniture factory; but he did not receive any reply. At the end of 2005, at a discussion meeting of retired cadres, the old secretary reported the situation regarding pollution and cancer in Dingbang to the Township Party Secretary, but he did not take a position either. Afterwards, some villagers used the 'mayor's hotline' to report the pollution situation in the village. The city instructed the district EPB to send people down to investigate and concluded that even though the metal furniture factory discharged pollution, the pollutants did not contain toxic substances, and were not directly related to the Dingbang villagers' cancer. When reporting to the village committee and government also proved to be in vain, Dingbang villagers approached the media for help and called the city TV station talk show. A reporter from the programme came to Xiqiao to conduct investigations, but afterwards the villagers did not see the programme on air. After enquiries they found that the programme did not get approval and could not be broadcast.

Unlike the gentle, reasonable negotiations of the Xiqiao villagers, Jiannan residents resorted to intense conflict in an attempt to make the polluting enterprise stop production or relocate. In September 2010, more than 50 Jiannan residents went to the glass factory and locked its front gate, blocking the factory vehicles from entering or leaving. In fact, since 1995, Jiannan residents persistently organized individuals to petition all levels of government. In December 2010, when we investigated, the enterprise was still in production. We learnt from the County EPB that even though the pollution could not be confirmed, judging from the situation that was developing, this glass factory would be relocated away from the residential area. Afterwards, we learnt from an interview with the local government leader responsible for environmental protection that a new location had already been found for the glass factory, and it was just a question of time. Furthermore, according to law and administrative procedures, there was no basis for immediately shutting down the factory as the residents wished.

The limitations of the strategy of attempting to eliminate the source of pollution are clear. The method is simple and effective, but in reality, it is difficult to accomplish, at least within a short period of time. A compromise solution that is comparatively effective and more feasible is to ensure enterprises emit only waste products that have been treated and only in quantities that are legal, and that they operate under the effective supervision of the local government, As the result of Xiqiao villagers' negotiations with the metal furniture factory, even though the pollution was not eliminated as they hoped, the enterprise did make an effort. For example, starting from 2003, the factory built pollution treatment facilities, and began treating its waste water. According to residents, even though the factory still discharged pollution secretly, after 2005 the situation improved. Additionally, in December 2015 we learnt from the district EPB that the district was in the process of planning and implementing a unified system for treating rural pollution sources. We also saw that central waste water treatment pipelines were being laid.

Migrating Away from the Source of Pollution

If pollution from the enterprise continues to affect residents, and they are powerless to drive the enterprise away, some residents who are in better economic circumstances will choose the exit strategy of migrating away from the pollution source. In fact, in the course of our research many people made statements that conveyed this intention such as, 'we can't afford to provoke the factory, but we can afford to escape it'. Even though it is frustrating, this is also an effective method to evade risk. But whether a person can really escape depends on their psychological and economic situation.

After Dingbang villagers went through all the effort of the 'three-way' approach and still did not achieve the expected results, they no longer had any great expectations of government. Some villagers who were in better circumstances attempted to 'escape' from their own village. At the time of our research, a third of the 25 households in Group 5 had already purchased properties on the private market in the town or city. Naturally, their financial situation and job requirements were the main reasons for moving out, but pollution was also an important consideration. Mr. Chen, a Dingbang villager, had stomach cancer eight years before and had been cured. He was convinced his cancer was directly related to pollution. He therefore lived in the city most of the time and did not want his granddaughter to return to the countryside to visit. The fact that 'the environment is so awful' was one of the reasons he was unwilling to live in the village. In contrast, not a single household in Xiqiao Group 5 had built a new house or renovated an old house within the traditional village boundary in the past ten years. Finances and transportation issues were the primary considerations in this decision, but the environment was also an important factor.

From a broader perspective, environmental pollution was also one of the factors causing the hollowing out that can be observed in villages in some regions in China. The flow of rural labour to the cities, especially higher skilled workers, has already become a common trend. Cities are attractive to the rural population in terms of employment, income, transportation, medical facilities, and educational institutions. The condition of the environment is now also an important reason why many village residents choose to move to the city. At the time of our research, and especially in the developed areas of eastern China, where the rural environment had been neglected for a long time while the urban environment was improving, villagers who were in a good financial situation often chose to live in the city, and this led to the further marginalization and hollowing out of villages.

Changing Water Sources

Even though attempting to completely eliminate the pollution source offers greater security, in reality it is difficult to accomplish. Moving away from their original home is also often restricted by villagers' employment, education, and financial situation, and not many households are able to move. The most feasible way to avoid or eliminate possible risks is therefore by changing one's own lifestyle and habits. Water is an important medium of exposure to pollutants, through drinking water and through food grown on polluted land. Abandoning polluted water sources (or water sources that are suspected to be polluted) is therefore a relatively safe and practical method to reduce risk. In fact, solving the problem of water usually becomes the main demand of environmental activists.

Shangba and other villages affected by mining pollution adopted this strategy. The water of Shangba and Liangqiao had been affected by B Mining Enterprise and other mining pollution since the 1970s. In order to avoid the health risks associated with water pollution, villagers initially dug wells so that they could consume ground water. Later, villagers discovered that there were also problems with the well water, so some of them raised funds themselves to bring in clean water from nearby mountain springs as drinking water. Some villagers even rode motorbikes to bring water back from the mountains. But the mountain spring water was limited and could not satisfy demand for irrigation water. Furthermore, villages located far from mountain springs had no way to source water in this way and so some villagers still had to continue to drink well water. Later, some smart villagers discovered the business potential inherent in this situation and started driving into the mountains to collect spring water to sell. In addition to polluted drinking water, irrigation of farmland was also a problem. Starting in the 1980s, Shangba villagers continuously organized a series of petitions and they considered this one of the most effective ways to solve the problem.[11] Shangba also took full advantage of the influence of the media. The CCTV programme Half an Hour on the Economy reported on mining pollution in Shangba twice, in 2001 and 2005.

11 The leaders of the 21 villager groups in Shangba were financed by their groups to go to the provincial capital to petition. Sometimes the Shangba Village Committee joined in petitions with other villages. Because B Mining Corporation is a provincial state-owned enterprise, the separation of economic interests means that although the County government is still under political pressure from higher levels of government, it still silently allows or supports the villagers' petitions. If petitioning has a 'result', B Mining Corporation will need to give pollution compensation to the County Government and residents, and so the County Government is a beneficiary of the petition. However, the County Government does not support villagers 'stirring up trouble' for the local mines. Officials from the County EPB once said to villagers that if they see wastewater from B Mining Corporation coming down the river, then they should call and inform them, but if it is pollution and waste water from the privately-owned mines then not to inform them. Thus, B Mining Corporation insists that if the responsibility for pollution is not shared between the state-owned mine and the privately-owned mines, then they will not be fully responsible for it. This is also an important reason why the issues of pollution control and compensation have not been completely settled.

In 2005, the *Nanfang Metropolis Daily* also reported on pollution in Shangba and some other villages. Foreign media and research institutions from countries such as England, Germany and Japan also went to Shangba and other villages on multiple occasions to conduct interviews and fieldwork. Shangba villagers also attempted to solve the problem through legal means, and the local government supported them in filing a lawsuit against B Mine. Through the relentless efforts of the villagers and other social actors, in 2008, more than 14 million yuan was provided by the province, the city, and B mining company to build the Shangba Reservoir, and tap water was made available to all households in the village. The issue was officially resolved, but when we conducted fieldwork, we found that for complex economic and social reasons, the drinking water problem had still not actually been solved.

Unlike Shangba, where a powerful clan organization played a role, Liangqiao was not given a 'special project' to solve its water problem. However, because the population of Liangqiao is relatively small, and it is situated upstream, it was easier to bring in clean mountain spring water, and the issue of drinking water could basically be solved by the villagers themselves. Yet, neither in the influential Shangba nor in relatively neglected Liangqiao had the problem of polluted irrigation water been resolved.

Jiannan residents fought for many years to get running water installed. As described earlier, several indicators in the two water samples they sent to the provincial department for inspection exceeded the standard. Even though it is still unclear why this was so, and the relationship between these indicators and cancer is even more uncertain, residents claimed that the water tests were proof of pollution from the glass factory, and that this was the cause of the high incidence of cancer. They therefore asked the enterprise and government to install running water. In fact, their calls for running water had started as early as 1992 and in 1995, 1999, and 2000, they petitioned multiple levels of government. Between 2000 and 2010, because the high incidence of cancer among residents attracted attention, their petitions became more frequent. In 2009, with partial funding from the government, and 300 yuan from each household, tap water was finally installed. From a rational point of view, the residents' demand for running water may not seem a big request, but in a relatively undeveloped region, local financial resources were limited, and it was not until 2009 when there was a road being built that they could take the opportunity to connect Jiannan water lines.

In conclusion, changing water sources is a plausible solution to avoid health risks from pollution. If the original water source has health hazards then these can be eliminated; and even if there is not a serious problem with the original source, the water quality will be improved, and the supply is more convenient. Water improvement projects undertaken by the government or NGOs take both of these concerns into account and are a relatively less precise approach to responding to pollution. However, even when it comes to improving water sources there are complex relationships in play that involve different interests, and which make the seemingly simple avoidance of health risks a complicated economic and social issue. For example, the reservoir that was built in Shangba at significant cost was unable to completely solve the drinking water issue for villagers due to problems with the quality of the construction. In Jiannan, again, because the pollution-health issue was enmeshed in the socio-economic context of the farm's transition out of state ownership, the situation was very complex. For example, the family of opinion leader Mr. Zhou claimed that his well water had been polluted and that this was causing the high incidence of cancer, but at the same time, even though they used tap water, his family also continued to use water from the well.

Changing Food Sources

Consumption of polluted food is a major health risk and substituting food that is contaminated (or suspected to be contaminated) is a relatively effective solution to this. On the face of it, drinking water would seem to be a personal matter, but in fact, getting a secure supply of water is mostly a public affair, and so efforts to address the issue of drinking water usually involve collective strategies. In contrast, the food problem is a personal issue, and the actions of individuals and families are more crucial. At the same time, the sale of contaminated agricultural products on the market also redistributes pollution, causing pollution and its health risks to be 'socialized' and evenly spread, and this once again turns a private issue into a public one.

Because the irrigation water had been polluted, there were excessive levels of heavy metal elements in Shangba's rice, vegetables and other produce. Some villagers had obtained information about this from scientists, but due to the cost, convenience, and other factors, the majority of villagers still ate rice and vegetables they planted themselves. In Dingbang, the villagers did not know exactly whether or not their grain and vegetables were polluted. They usually made judgements about whether these foods were safe to eat based on their personal experience. Out of concern for safety, some villagers chose to buy rice from industrially underdeveloped areas such as northern

Jiangsu or Anhui, and a few others used well water instead of the polluted river water to irrigate their vegetable patches.

Selling agricultural products believed to be unsafe was another common practice adopted by residents of 'cancer villages', but due to their notoriety, especially in cases like Shangba where rice contaminated with heavy metals had been widely reported by the media, it was hard to sell their crops to others. Dingbang's status as a 'cancer village' was not as well known as Shangba's, but perhaps due to differences in regional social culture, the people near Xiqiao were more sensitive. In the markets there, once the subject of cancer came up the mood changed immediately, and no locals were willing to purchase vegetables sold by people from Dingbang.

The export of sugar cane and other crops outside the region dispersed and homogenized risks from contaminated food. For economic reasons, Shangba villagers had switched in the years before our research to planting sugar cane, which was mainly purchased by external merchants and sold to consumers outside the province. This not only lowered the local residents' health risks, but also reduced economic losses caused by pollution. Hypothetically, in the past, if a farming household had four members, and three *mu* of land, and all the agricultural products from that land were consumed by the household members, then the risk from heavy metals in the crops would be very high, but the scale would be limited to that household. But when sugar cane from the same three *mu* of land was sold outside the area, if we assume that 10,000 people consumed it, then the risk from heavy metals would be evenly distributed across that much larger population, with a much lower average risk. However, although when risk is spread in this way there will not be an obvious impact on health in the short term, over a longer period, if this logic is extended to the whole food chain, the quantity of pollutants in all the various fruits, vegetables, grains and animal products we eat would also increase, leading to a greater health risk on a societal level. The fact that monitoring of serious diseases by the Jiaxing City CDC showed that the cancer mortality rate within the city was rather high overall, despite showing no particular geographical characteristics, is a good footnote to this point.

On a related note, the stigma of being known as a 'cancer village' had unexpected consequences for villagers. For example, media attention brought hope and some benefits for the Shangba villagers in terms of solving the problem, but it also revealed to the outside world that Shangba was severely polluted and had a high cancer mortality rate. At first, in order to gain outside attention, sympathy and support, Shangba villagers described in detail how they were affected by pollution and the tragic situation with cancer. But stigma also brought problems. Apart from the difficulties described above in selling agricultural products, the marriage prospects of the village's young people were also affected. In fact, it is common for disease to trigger this kind of unintended social consequences. A similar thing happened in Japan. Before Minamata disease was clearly diagnosed, people living around the area excluded and discriminated against victims of the disease out of fear of infection. More recently, in China, people from 'AIDS villages' were also shunned and discriminated against by the public. When village cadres in Shangba refused to provide researchers with a list of cancer victims, this was also an attempt to reduce outside attention and remove the stigma from the village.

Discussion

Determining the relationship between cancer and pollution is an incredibly complex task that is in many cases impossible even for natural and medical scientists. It is even more difficult for ordinary villagers. But their awareness of and responses to the relationship between cancer and pollution is not only subject to the limitations of science. Their perceptions and actions are also heavily influenced by economic and social factors.

Economic factors have different dimensions. Financial resources are the primary consideration in whether a person is willing or capable of moving away from a polluted area and whether an enterprise can be relocated. When the relationship between pollution and cancer is not confirmed or when there are controversies, villagers can still use the situation as the basis for fighting for access to running water and other public utilities. Meanwhile, local governments may do their utmost to protect local enterprises and act as their partners in crime, or they may fight side by side with villagers who are affected by pollution and take part in the division of the spoils. Both positions are informed by economic considerations. Furthermore, the suspicion of pollution-related-cancer is sometimes used as an excuse to start construction projects, foster corruption, and spend tremendous amounts of money to build substandard reservoirs. Although the tactic of selling contaminated agricultural products grown in 'cancer villages' to outsiders minimizes the financial loss of the villagers, it also allows pollution to spread throughout society, illustrating Beck's notion of the new, democratic nature of risk in modern society.[12]

12 Beck, *Risk Society: Towards a New Modernity.*

How does the social structure of the countryside affect villagers' responsive strategies?[13] The situation with pollution in Shangba and Liangqiao was very similar, but due to differences in lineage and other aspects of social structure, they produced very different patterns of information dissemination and social action regarding the cancer-pollution relationship. Of the five villages that were quite heavily polluted by B Mine, Shangba had a larger population of the same ancestry, and its clan organization had strong organizational capacity. This enabled the villagers to be active in organizing petitions and attracting media and NGOs to join their efforts, drawing widespread attention from the outside world. Shangba was a clan settlement where most of the villagers shared the surname *He*. The clan's ancestral hall stands near the village committee office and as we observed in our fieldwork, the power of the clan influenced many aspects of life in the village. First, in accordance with tradition, it served as a legitimate and reliable authority that played a decisive role in uniting and coordinating the villagers' ideas, balancing economic interests, and organizing group actions. Second, villagers of same clan have a very strong sense of belonging to the ancestral lineage, and this made it easier to for them to come together around public issues, forming a united front. Additionally, clan organizations are often interconnected with administrative institutions, or are the same as the village committee, and can have networks that extend further, to the township or even higher levels of government. This was the case in Shangba, where clan members made use of such connections.

Liangqiao is situated upstream and is nearest to the mines. It was not only affected by waste water, its fields were also covered by sludge which affected agricultural production. But the actual level of attention it received was far less than that given to Shangba. This was mostly because Liangqiao's population was smaller, and it was a village with mixed surnames formed by migrants who moved into the region at different times. In terms of ability to deal with facing public issues affairs, the organizational skills and resources of Liangqiao were clearly lower than those of Shangba.

The role of 'gentry power' (*shenquan*), which is a local tradition in the Jiangnan region, is also evident in the case of the retired secretary coordinating Xiqiao's response to the 'cancer-pollution' problem. In ancient China, imperial power did not reach below the county level, and many public affairs in the county were mediated by government officials who

13 One related subject is the research concerning information dissemination mechanisms and health risk resolving strategies in a village 'familiar society', see the next chapter.

had retired to the countryside.[14] But in fact, the condition for the operation of 'gentry power' is that there must be members of the 'gentry' who have returned to the countryside. Southern Jiangsu seems to have a particularly good foundation in this respect. In the case of Xiqiao, the original village Party Secretary had retired there, and he was familiar with both the actual situation within the village, and with the operational mode and logic of officialdom. These qualities, in addition to his pre-existing reputation and relative detachment from practical interests, made him an important figure in attempts to solve the pollution problem. Although this retired secretary did not seem to have played the role he should have done in terms of resolving this specific problem, some questions are still worth pondering. For example, why did Xiqiao villagers not cause much trouble? And from a broader perspective, looking at the entire region of southern Jiangsu and northern Zhejiang, why, even though the water pollution is incredibly severe, do conflicts between officials and the people still appear on the surface to have been less serious than in other regions and why did serious conflict usually occur in places with less gentry power? Why did enterprises entering the region later restrain themselves, and why did local governments make efforts to solve the problem in a holistic fashion? The role of 'gentry power' and approaches to the problem of pollution and cancer appear to be topics that require further research.

Unlike the gentry power in Xiqiao, in Jiannan we saw faint traces of the work unit system. This was an important organizational form during the period of the planned economy after the founding of the PRC. Although it incorporated certain aspects that were inherited from earlier forms of social organization in China, the work unit was also unique in many ways. When housing or social welfare were being allocated, individuals seeking to obtain benefits outside or beyond the standard provisions would use all kinds of tactics to cajole and pester unit leaders. Some would stop at nothing, and sought out the unit leaders every day until they got what they wanted. Similarly, in their unrelenting efforts to solve the 'pollution-cancer' problem, some residents of Jiannan could not take an objective view of differing opinions on the relationship between pollution and cancer. They remained convinced of their own views and fought for their interests in this same kind of way. In fact, the 'pollution-cancer' issue is only one problem that Jiannan residents kept petitioning about; there were also ongoing actions to fight for various social benefits. It seems that this psychology also affected the logic of their actions in responding to pollution.

14 Wu, Fei, et. al., *Imperial Power and Gentry Power*, 1988.

Works Cited

Beck, Ulrich. *Risk Society: Towards a New Modernity.* London: SAGE Publications, 1992.

Chen, Ajiang. 'On Harmony between People and Water.,*Journal of Hohai University (Philosophy and Social Sciences Edition)* 2008 (issue 4): 25-30, 100-101.

—. *Secondary Anxiety: A Social Interpretation of Water Pollution in Lake Tai.* Beijing: China Social Sciences Press, 2010.

Ding, Fengchu. 'On Foreign Environmental Rights Infringement Causality Relationship Theory – Improving China's Related Theories'. *Social Science Research* 2007 (issue 2): 96-100.

Funabashi, Harutoshi. 'Minamata Disease and Environmental Governance'. *International Journal of Japanese Sociology* 15, no.1 (November 2006): 7-25.

Liu, Yisheng, Gao Yi, Wang Kangwei, Mai Xiaohan, Chen Guangdao, and Xu Tongwen. 'Etiological Research on Malignant Tumours of the Digestive System in a High Incidence Village in Guangdong'. *China Tropical Medicine* 2005 (issue 5): 1139-1141.

Lora-Wainwright, Anna. 'An Anthropology of "Cancer Villages": villagers' perceptions and the politics of responsibility'. *Journal of Contemporary China* (2010), 19(63):79-99.

Wu, Han, Fei Xiaotong, et. al., *Imperial Power and Gentry Power.* Tianjin: Tianjin People's Publishing House, 1988.

Zou, Xiaojin, Qiu Rongliang, Zhou Xiaoyong, et al. 'Research of Health Hazards of Dabao Mountain Mining Area Heavy Metal Pollution'. *Environmental Science Daily* 2008 (issue 7): 1406-1412.

10 Villagers Strategies for Mitigating Environmental Health Risks[1]

Chen Ajiang and Cheng Pengli

Abstract

The unclear scientific relationship between pollution and cancer, combined with weak governance systems, means that villagers have only incomplete information about pollution and disease. Their own lack of scientific knowledge is a further obstacle to their accurate perception of health risks. But villagers do not just resign themselves to their fate. They use the social structure and cultural mechanisms of the 'familiar society', to maximize their use of available information, and they rely on their life experience and common-sense strategies to mitigate the impact of pollution. Village social structure also affects the actions villagers take to avoid pollution. Thus, although they lack scientific guidance in mitigating health risks from pollution, their actions are informed by their own unique empirical logic.

Keywords: information circulation, local knowledge, social structure, Tai Chi, subjective agency

Over the last 30 years, rapid economic development and industrialization have caused environmental pollution and ecological degradation in China, with many negative consequences for public health. The frequent appearance of media reports about 'cancer villages', and 'villages with strange illnesses' indicates the severity of environmental health risks and reveals the high degree of public concern about them. This has become a pressing social problem.

1 The original article was published in *Journal of Nanjing Agricultural University* (Social Science Edition), 2011 (issue 2).

Chen Ajiang, Cheng Pengli, and Luo Yajuan, *Chinese "Cancer Villages": Rural Development, Environmental Change and Public Health.* Taylor & Francis Group 2020
DOI: 10.5117/9789089647221_CH10

Environmental pollution not only has big health effects, but also heavy social and economic impacts. The classic case is that of Minamata disease in Japan, which affected a total of 37,360 people, of whom more than 1,000 died. By May 1997, the issue of compensation for Minamata disease victims had basically been resolved, but the social problems it caused had not.[2] In China, there have been numerous instances of environmentally-related disease, including the widely reported appearance of multiple 'cancer villages' along the Shaying River, the largest tributary of the Huai,[3] and high rates of cancer in a village within the jurisdiction of Yancheng City, Jiangsu that suffered severe air and water pollution.[4] High levels of lead in the blood of residents of Fengxiang in Shanxi, and the arsenic contamination incident in Yangzong Lake in Yunnan are other examples. These incidents are sufficient evidence of the severity of environmental health risks in China.

At the same time, it is difficult for ordinary villagers to become aware of these risks. Their limited awareness is partly due to the underlying limitations of scientific knowledge about the relationship between pollution and disease, but it is also related to social structure and institutional arrangements. Conducting biomedical and epidemiological research into the relationship between environment and health continues to be challenging. The existence of multiple cause and effect relationships leads to a lot of disagreements and it is hard to produce definitive results. The uncertainty of the medical science in turn makes it difficult to assess the status of legal evidence and as a result many 'cancer villages' and 'villages with abnormal illnesses' remain just reports or rumours. It is hard to find evidence that would be recognized in a court of law. This is ultimately the reason why it is usually hard to win environmental law suits in China.[5]

Their relatively low level of education and scientific literacy mean that it is very difficult for villagers to gain a full and objective grasp of the complex, uncertain relationship between environmental pollution and human health. On top of this, people living in polluted rural areas face many obstacles when it comes to obtaining information. Polluting enterprises often deliberately withhold information about emissions; and delays and lack of transparency in information disclosure by government departments hamper villagers' ability to judge the nature and extent of environmental health risks. Yet in spite of these obstacles, ordinary people are not as

2 Funabashi, 'Minamata Disease and Environmental Governance', pp. 7-25.

3 Li, 'Huai River Tributaries: Pollution Creates a Number of "Cancer Villages"', p. 1.

4 Ye, 'Killer GDP: Yan City Cancer Village Industrial Pollution Investigation', p. 3.

5 See Chapter 3 of this book.

completely helpless in the face of environmental health risks as one might think. In fact, they manage to deploy a number of quite effective strategies to avoid health risks, drawing on their own circumstances, life experience and practical wisdom.

This chapter draws on earlier empirical work to explore these strategies. One case we discuss is Xu Village, which is the hometown of one of the authors. It is a natural village of 28 households and around 110 people in Wujiang, Jiangsu, close to Zhejiang. Xu Village has served as his long-term observation site, first for understanding rural industrial development and more recently for researching the problem of water pollution. In the summer of 2010, he focused specifically on the issue of environment and health among villagers over 30 years old. The second source is our research on Xiqiao, an administrative village in northern Zhejiang, near Jiangsu. Villager Groups 4 and 5 (which make up the natural village of Dingbang) had been termed a 'cancer village' due to the high incidence of cancer caused by industrial pollution. Starting in 2005, we conducted four rounds of environmental sociology research in the village and in December 2010 we conducted a survey, in depth interviews, and other fieldwork on the topic of 'Perceptions and Responsive Strategies to Environmental Health Risks'. The third source is our December 2010 study of 'awareness of environmental health risks and counter strategies' for which we conducted fieldwork in Jiannan Village in Shangrao, Jiangxi, and Shangba Village and Liangqiao Village in Shaoguan, Guangdong. We also draw on insights that we and other members of the research group have formed while conducting earlier research on the relationship between people and water in Northern Jiangsu, Eastern Henan, and similar places. This chapter seeks to go beyond case studies and use this body of past research to make a broader analysis of how rural dwellers respond to and mitigate health risks stemming from the environment, despite their limited knowledge.

Awareness that Pollution Causes Disease and Information Barriers

In rural areas where environmental pollution has caused a high incidence of disease, the fundamental obstacles to ordinary villagers' awareness are the underlying scientific uncertainty of the pollution-disease relationship and the incompleteness of the information available. Villagers may think the water is a strange colour or has an odd taste, but they will find it very hard to find out what is causing this. They would need to figure out who is discharging waste water, what kinds of pollutants are in the water and in what concentration, what kind of health risks these pollutants present and what kind of harm they might suffer if they drink the water, use it for irrigation, or eat fish from the river.

Some of these questions may involve scientific uncertainties, but even where there are definitive scientific findings, villagers' understanding is limited because this is specialized knowledge and the situation is also often very complex. A great many different pollutants may enter a given body of water from multiple industrial enterprises producing all kinds of products using different techniques. Some pollutants may be very colourful, but not very toxic; while others may have no strong colour, but be extremely dangerous. Once pollutants enter the water, they may also undergo chemical interactions and biochemical changes. It is therefore very difficult to know which pollutants will harm health and the exact level of hazard they present.[6] For example, clearly identifying and confirming methylmercury as the cause of disease in Minamata, Japan became a famous case in the history of medical research on pollution-related diseases. Yet even in this case, where there was a relatively clear relationship between pollution and disease, it took a long and complicated process for the results to be finally accepted by the court, and cases in which there is a clear scientific relationship between pollution and disease are incredibly rare. In the majority of cases, it is not clear at all.

In addition to the innate complexity and uncertainty of science, the lack of transparency in information about environmental pollution is also a big problem. It is very common for enterprises to discharge excess levels of pollution or make secret emissions to lower their costs and maximize profits, and enterprises that do this naturally do not release accurate information about pollution. In these situations, the local government usually cooperates with the enterprise. For example, in the case of Dongjing Village discussed in Chapter 3 of this book, the Julong Chemical Plant was established as part of the local government's campaign to attract investment. When construction of the enterprise started, township and village cadres and representatives from the chemical plant promised the villagers that the things produced by the chemical plant were cold capsules and other chemical products for use in daily life which would not cause pollution. They also promised that the plant would treat the water with a recycling-reuse technology, so that not one drop would be spilled outside, and there would not be any pollution to the surrounding water sources and air. However, villagers very quickly found that the rice they grew tasted of chemicals produced by the plant, and they did not dare to eat it anymore.[7]

6 Chen, *Secondary Anxiety: A Social Interpretation of Pollution in the Lake Tai Basin*, p. 75.

7 Please see Chapter 3 of this book.

Institutional development is also weak. The government has still not established an open and transparent information disclosure system that specifies the responsibilities of the government departments concerned with regard to disclosing information to the public in a timely fashion. Some local governments do not disclose the information they do have because they have a stake in the polluting enterprises, but it is even more common that institutional development lags behind the level of economic development in the area and therefore behind the public's need for information that might enable them to protect themselves from harm. The controversial issue of waste incineration power plants is an example. The pro-incineration lobby believes that waste incineration technology is mature, and often cites Japan and Germany as examples that the risk can be controlled. But the anti-incineration lobby points out that China's environmental pollution information and regulatory system cannot be compared to Japan or Germany. During our own recent research on the issue we found that there are only a few organizations in China capable of testing for dioxin, and it is very expensive. As a result, even the agency that is responsible for regulating and controlling dioxin knows very little about the actual situation regarding emissions. The regulatory system itself still does not have sufficient capacity, not to mention ordinary people.

Although the official pollution-disease information disclosure system is inadequate, the development of the media and growing social diversity mean that villagers can still obtain some information through other channels. Television and other media are pervasive in the countryside and very influential, so villagers can often get some information and knowledge from this source. For example, the China Central Television (CCTV) programmes 'Focal Interviews' and 'News Investigation' and other similar shows have investigated a lot of environmental health cases and spread a lot of scientific knowledge. When we were conducting fieldwork in villages, we often heard people cite certain programmes from CCTV as authoritative sources, to clarify or explain the local situation.

Information Sharing in the 'Familiar Society'

Many factors affect the public's awareness of risks and responses to them, and the possession and communication of information are crucial. In rural communities, because of the influence of community structure and local culture, the way in which villagers take in and process information about environmental health risks is deeply influenced by the fact that they live in a 'familiar society' (*shuxi de shehui*) and this works to reduce environmental health risks to a certain degree.

The 'familiar society' is a concept coined by Fei Xiaotong in his book *From the Soil*. He argued that an important characteristic of Chinese rural communities is their localness. 'Living in a rural society is full of localness. Localness refers to the geographical limitations on their range of activities. There is relatively little contact between communities, and they are isolated from one another, with each maintaining its own separate social circle'. As the result of the constraints of localness, people within the village itself are very familiar with each other. The village is therefore a 'familiar' society, and 'a society without strangers', in which communication between villagers does not require many words or written language. This familiarity is the result of habituation, and 'society and individuals are intimately connected'.[8]

The 'familiar society' is one in which communication is very good. Even though not much information is available, it is very transparent, it travels fast, and it is widely shared. No household has a secret that others do not know. It can be said that in rural society information is limited but communication is unlimited. Chinese rural communities have already undergone enormous changes, and they no longer conform to the ideal type of the 'familiar society', but in terms of the way in which information is communicated within villages, these characteristics are still present. This is in strong contrast with the 'society of strangers' that is typical of urban communities. Even though they may live in the same apartment building their whole lives, urbanites may never get to know their neighbours, and they may not even know the name of the person who lives next door.

In the 'familiar society', information, and especially information about incidents related to public safety such as fires, or affairs that involve all villagers, spreads incredibly fast, is widely shared and highly transparent. Villagers mull over such incidents, discussing them repeatedly, to the point where they practically know them by heart. For example, after news reached Xu Village that the Zhang family's uncle was in a car accident on a certain road, and had been sent to the hospital, villagers passed it onto each other, until almost every villager knew all about the Zhang family's uncle being hospitalized. The incident was repeatedly brought up and discussed throughout entire summer and it was also connected with later accidents on the same road. After repeatedly speculation and deductions, the villagers finally came up with their own interpretation of the event:

8 Fei, *From the Soil*, pp. 41-43.

> The new road was built too well, and the cars drive too fast, especially the ones from outside the area [...] Elderly people on three-wheelers should always get off and push them across the road, even if it's a bit slower.

By the end, 'a person on a three-wheel vehicle should push it across the road, even if it's a bit slower', and other key information about the incident, was known by almost every household, and gradually became the villagers' shared knowledge about how to deal with traffic risks. In rural areas where industrial pollution is severe, information about it spreads in a similar way. For example, some called Guan who works at a chemical plant discovers a tumour in his stomach, and when he goes to the hospital for an examination, the doctor says it requires surgery. The information about this then spreads through the village, and it is repeatedly brought up and discussed. In the end, almost every villager knows about Guan going to hospital and having an operation and all the speculations about whether his working at the chemical plant had something to do with his getting the disease. And they will engage in all kinds of discussions that expand upon the topic. For example, if his illness might have something to do with his working at the plant, they will debate whether when he recovers should he still continue to work there or change jobs, and whether other workers in similar chemical plants should continue to work there or make changes.

In brief, in a traditional 'familiar society', information from the outside world is limited, but the degree to which it is shared is high, and this unlimited use of limited information helps villagers to avoid environmental health risks to a certain extent. Of course, with economic and social development in rural areas, the traditional 'familiar society' has already undergone many changes, and rural communities have become more differentiated. On the one hand, this is evident in the differences brought by uneven economic development across regions. It is also evident in differences between groups within the same rural community. In regions where the economy is more developed and society is more open, rural communities have gradually become less different from urban ones. The spread of televisions, phones, the internet, and other public media is constantly increasing the amount of information rural communities receive and diversifying the channels through which it comes. In economically less developed regions, even though rural communities are also gradually becoming more open, overall progress is still relatively slow.

Furthermore, even within the same community, differences between villagers were also relatively large, for example in terms of their level of education and life experience. Age also plays a role. When new information enters a village, young people's methods of distributing it are influenced mostly by the modern technology, but middle-aged and elderly people basically still spread information in conformity with the old habits of the 'familiar society'.

Ways of Mitigating Health Risks

Most environmental health risks in the Chinese countryside today are the result of pollution coming from the 'unknown world' outside. In traditional society, villagers can deal with all kinds of public affairs easily, partly because information is transparent, and also because the village community has already formed a set of relatively stable and effective rules and procedures. For example, when faced with an act like littering that breaches the rules, the village will develop a set of strict and effective methods to deal with it, so that even though some people do not totally comply, the overall social goal is achieved. But when the community is confronted with industrial pollution from outside the village, these internal rules are no longer effective in dealing with it.[9]

However, even when it is difficult to stop or eliminate pollution from outside sources, villagers do not turn a blind eye to the health risks. In fact, they are very active, and their exploration and experiments produce a set of practical strategies to avoid harm. We refer to these as a form of 'mitigation' (*huajue*). When villagers become aware that pollution can damage their health, they draw on their life experience and existing knowledge to reduce the risk as far as possible and minimize the harm pollution does to their health. The original meaning of the Chinese word 'huajue' was to 'eliminate' or 'remove'. In Tai Chi, 'hua' has a special meaning. It refers to winning through ingenuity when engaging in the practice of *Tuishou*.[10] When your hands are pressed against each other and you are moving around, as your opponent's external force comes towards you, you neither wait passively, nor counter-attack aggressively. Instead you evade the pressure and make your opponent lose his purchase, and then you look for a chance to push back. In a similar way, villagers' strategies to 'mitigate' the health risks of pollution are neither a simple evasion, nor are they direct, active or radical resistance. Instead they allow the oncoming force to take its course and work around it. It may appear that they are adopting an attitude of resignation in

9 Chen, 'A Sociological Explanation of Water Pollution – A Case Study of Dongcun Village', pp. 62-69.

10 *Tuishou* is an exercise in Tai Chi in which two people take part.

the face of adversity, but in fact, although there is some implied resignation, while villagers allow the pollution to take its course, they make an effort to mitigate the effects and reduce the health risks causes to a minimum.

When they lack scientific knowledge, villagers use their common sense to avoid risks. Our common sense tells us that if we consume toxic water or food, we might get sick, so the method villagers most often use is not to consume food or water that is polluted, or to consume less of it. When villagers find the river water has a strange odour, they switch to well water for drinking and only use the river for washing clothes or vegetables. More cautious villagers will use well water to rinse vegetables that they have already washed in the river. And when they find that the river fish have a strange smell, they will try not to eat fish and substitute other foods. To reduce the physical damage from pollution to a minimum, they will usually sell grain and vegetables that are considered risky. The villagers of Dongjing and Xiqiao villages did not eat their own rice and vegetables because they suspected they contained a lot of pollutants and they had a pungent smell. Some villagers worried that the polluted river water might affect their vegetables and so they used well water or tap water for irrigation to reduce the risk.

Villagers have also figured out their own individual ways to evade health risks based on their experience. In the summer of 2010, when staying in Xu Village for a short period, the author noticed that several families had planted arrowhead plants[11] in earthenware jars. Curious, he chatted with them about it and discovered a whole other set of secrets. The following discussion shows how some detail-oriented people in rural communities are paying attention to health issues.

> Question: How many arrowheads can you grow in an earthenware jar? [The subtext to this question is that these jars are usually only used in the city where there is no land, but there are lots of ponds and ditches in the countryside, so why is this 'miniature pond' necessary?]
> Answer: Every jar can produce at least half to one kilogram of arrowhead! I have so many jars, and if I grow five kilos it will be enough for our family.
> Q: Then why not plant it in the ditches? You only need to plant a little for it to be enough for your family.
> A: That is different, the ditch is not clean.
> Q: Why?
> A: You didn't see? The water used for washing things by the well, washing powder, washing detergent, and all those chemicals flow into that ditch.

11 This is a kind of plant that grows in water.

According to Fei Xiaotong, the traditional social structure in China has a 'differential mode of association' (*chaxugeju*).[12] And the kinship and local relationships in rural society do indeed still influence and constrain the way villagers handle things, including the way in which they avoid environmental health risks. We noticed that, typically, parents do not allow young people to work in factories with more serious pollution, not only out of concern for the young people themselves, but also out of concern for the next generation. If a child from a certain family plans to work in a chemical plant, their parents will carefully weigh things up first. If the parents are not very wise, their close relatives and friends will warn them, and when there really is nobody to warn them, there are always people in the village who like to 'mind other people's business'. When these people are chatting, they will say, 'It's not a good deal for so and so's child to go and work at a chemical plant. Even if the salary is high it is not worth it, because you can't know that it won't end badly.' For example, old Jin in Xu Village always worked at a chemical plant, and he had a good income. His son followed in his footsteps and worked in the same plant but after a while, even though the wages were higher, Jin Junior gave up that job and went to a textile factory with average wages and harsher work, and the main reason was to avoid health risks.

Aside from the kinship and local relationship structures of rural society, the way in which the relationship between elites and commoners are structured also influences the way in which villagers avoid environmental health risks. Previous research has found that elite adoption of modern technologies is key to its dissemination in rural societies. Health knowledge and avoidance of health risks also follow this logic of dissemination from the elite to the masses. 'Barefoot doctors' who were trained during the People's Commune period had a huge influence on villagers. Even though they did not have a very strong grasp of medical theory, because they lived in the countryside, they understood rural society. They were familiar with the situation of the peasants, and they had in depth observation of the reality of rural life. Furthermore, they were also able to use language the villagers could understand to explain things they cared about. For these reasons, the explanations of barefoot doctors often become the accepted word in the village. For example, a barefoot doctor once advised a relative who loves eating river clams not to eat them in the future, because the toxins from the chemical plant had been absorbed and accumulated in the clams. Statements like this can easily spread within the village, and gradually influence and

12 Fei, *From the Soil*, pp. 60-70.

modify the behaviour of ordinary people. At a certain point, nobody in the village dared to eat river clams.

In fact, industrialization and urbanization mean that in addition to the elite-commoner distinction, intergenerational differentiation is also becoming more and more apparent in rural society. The difference between young people and the middle aged and elderly is particularly striking. Because young people generally have a higher level of education, and they have more contact with the outside world, they can easily accept and use new technology. They can also quickly become aware of and understand issues related to environmental pollution and health. Older people are usually introduced to new technology by the younger generation, but older people are better at utilizing local knowledge and solving problems by figuring things on the basis of their experience.

In analysing villagers' ways of responding to environmental health risks, we found that because of the lack of institutional support, they were forced to figure out strategies by drawing on their own practical experience. The limitations of this approach are very apparent; the villagers do their best to mitigate environmental health risks, but they cannot fundamentally solve them. Their strategies are based on a mixture of science and blind spots and they have only a few reliable sources of information, including local doctors, elites, and television and newspapers. The strategies they adopt are therefore effective only up to a certain point. Given the high incidence of environmental health problems in rural areas, the government still basically needs to implement effective environmental management. At the same time, it is crucial to strengthen public education suited to the needs of rural communities to increase villagers' awareness of environmental health, and improve their capacity to respond to environmental health risks.

Conclusion

How do residents of rural communities mitigate external environmental health risks? We have shown that it cannot be simply assumed that villagers are completely passive and defenceless when facing modern industrial pollution. In dealing with these risks, they have their own practical logic. Their understanding of environmental health risks and their strategies for avoiding them are not based on the kind of logic that 'outsiders' or most people see on the surface. They often originate from their daily life experience; a form of history manifested in the person. On the other hand, these strategies are also influenced by the local environment of the 'familiar society' in which villagers live. To understand this kind of practical logic, researchers need to start from the perspective of the villagers' actual lives, and extend their analysis through a process of targeted participatory observation.

Environmental health information and knowledge are the prerequisites for villagers to respond to environmental health risks. Industrial pollution and the diseases associated with it are very foreign to villagers. Enterprises deliberately conceal information about pollution and the government's system for information disclosure is imperfect, and this raises the awareness barrier for villagers as they seek to respond to environmental health risks. However, within villages, membership of the community is very stable, and interactions between villagers are also very frequent, forming a 'familiar society' that has quite a lot of social resources. When facing a common external risk, villagers make full use of the social networks of this 'familiar society'. Scant information about pollution spreads quickly and is repeatedly mulled over, with the result that limited information is used in an unlimited way to reduce health risks.

In discussing the opposition between modern technology and everyday knowledge and experience, postmodernist scholars have engaged in a great deal of debate and reflection. The anthropologist Clifford Geertz advocated 'thick description' that gives special weight to local knowledge and he went so far as to discuss everyday knowledge, religion, science and ideology on an equal basis.[13] When faced with the possibility that pollution may cause harm and scientific knowledge is lacking, local knowledge and daily life experience can be somewhat useful. When villagers realize that a water source or food may have been contaminated, they draw on their experience and do their best to avoid it. These experiences are already internalized, and people automatically adjust their own behaviour and way of life. For example, when some pioneering villagers started growing arrowhead plants in ceramic jars, other villagers imitated them. The innovative practices of community elites not only solve their own problems, but will quickly influence and spread among ordinary people.

Works Cited

Chen, Ajiang. 'A Sociological Explanation of Water Pollution – A Case Study of Dongcun Village'. *Journal of Nanjing Normal University* (*Sociology Edition*) 2000 (issue 1): 62-69.

—. *Secondary Anxiety: A Social Interpretation of Pollution in the Lake Tai Basin.* Beijing: China Social Sciences Press, 2010.

13 Geertz, *Local Knowledge*, p. 96.

Fei, Xiaotong. *From the Soil.* Berkeley: University of California Press. 1992.

Funabashi, Harutoshi. 'Minamata Disease and Environmental Governance'. *International Journal of Japanese Sociology* 15, no. 1 (November 2006): 7-25.

Geertz, Clifford. *Local Knowledge.* Translated by Wang Hailong et al. Beijing: Central Compilation & Translation Press, 2000.

Li, Junde. 'Huai River Tributaries: Pollution Creates a Number of "Cancer Villages"'. *Xinhua Daily Telegraph*, 29 September 2004, p. 1.

Ye, Wentian. 'Killer GDP: Yan City Cancer Village Industrial Pollution Investigation'. *China Business*, 24 April 2006, p. 3.

Index

For Product Safety Concerns and Information please contact our
EU representative GPSR@taylorandfrancis.com Taylor & Francis
Verlag GmbH, Kaufingerstraße 24, 80331 München, Germany